GEOMETRY AND THE IMAGINATION

The Imaginative Treatment of
Geometry in Waldorf Education

by

A. Renwick Sheen, M.Sc. (Tech.)

David Mitchell, Editor

"The moving power of mathematical invention
is not reasoning but imagination."
—A. de Morgan

Published by:
AWSNA Publications
The Association of Waldorf Schools of North America
3911 Banister Road
Fair Oaks, CA 95628

© 1991, 2002 by Stephen Sheen, "Lorien," Dale Rd., Forest Row, Sussex, England RH185
and AWSNA Publications, The Association of Waldorf Schools of North America

2002, fifth printing - completely reedited and revised

Library of Congress Cataloguing-in-Publication Data

Title: **GEOMETRY AND THE IMAGINATION**
   The Imaginative Treatment of Geometry in Waldorf Education

Author: Sheen, A. Renwick

ISBN # 0-9623978-2-2

All rights reserved. No part of this book, except for brief excerpts, may be reproduced in any form without written permission from the copyright holders.

Printed in the United States of America.

Editor: David Mitchell

Illustrations: A. Renwick Sheen and David Mitchell (after Sheen)

Proofreaders: Judith Grumstrup-Scott, Ann Erwin

Contents

List of Figures, Plates, and Tables 7
Editor's Comments 17
Preface 19

Chapter 1   Introduction 21
Chapter 2   Pre-Geometry 27
Chapter 3   First Lessons in Geometry 33
Chapter 4   The Pentagon and Pentagram,
            and the Golden Ratio 71
Chapter 5   The Four Rules of Arithmetic 111
Chapter 6   The Five Regular Solids 125
Chapter 7   The Conic Sections 141
Chapter 8   Projective Geometry 181

Postscript 247
Endnotes 249
List of Selected Books 251
Index 255

List of Figures, Plates, and Tables

Figure   **Page**

- *1*   Symmetry drawings by children of ages 7 to 9 years 29
- *1ab*   Relation of number and form 30

CIRCLES

- *2*   Family of concentric circles 36
- *3*   Family of circles with common tangent 36
- *4*   Fundamental circle design: six-division of circle 36
- *5*   Circle design involving a repetition of Fig. 3 37
- *6*   Circle design involving a repetition of Fig. 4 37
- *7*   The circling of a circle 37

TRIANGLES AND TRIANGLES INSCRIBED IN CIRCLES

- *8*   Family of isosceles triangles developed from equilateral triangle 38
- *9*   Family of scalene triangles developed from equilateral triangle 38
- *10*   Triangle design based on Figs. 8 and 9 39
- *10a*   Acute, obtuse and right-angle 40
- *11*   Family of equilateral triangles, sides halved, areas quartered 41
- *12*   Family of right-angled triangles, angle in semi-circle is a right angle 42
- *13*   Alternate angles equal and corresponding angles equal between parallel lines and a transversal. Also the exterior angle of a triangle equals the sum of the two interior opposite angles 42
- *14*   The three angles of a triangle are together equal to two right angles 42
- *15*   Equilateral triangle divided into six right-angled triangles: example of threefold symmetry 43
- *16*   Movement of equilateral triangle towards and beyond the center of a circle   43
- *17*   Interlaced equilateral triangles in a circle 43
- *18*   A modification of Fig. 17 43
- *19*   A 24-sided polygon with all its diagonals in a circle 44

SQUARES, RECTANGLES, AND RHOMBUSES

20	Family of squares, sides halved, areas quartered, square "begets" square	45
21	Star design based on square, octagon appears	45
22	Interpenetrating squares	46
23	Interpenetrating squares arranged in a circle	46
24	Star design based on two squares (or octagon) in a circle	46
25	Interlaced squares in a circle	46
26	Family of rectangles arising from a square, all of equal area	47
27	Family of rhombuses arising from a square, all of equal area	48
28	Family of rhombuses arising from a square, all sides of equal length	49
29	Design of rhombuses arising from a square	49

PARALLELOGRAMS AND TRAPEZIUMS

30	Family of parallelograms on same base and between same parallels as parent square	51
31	Family of trapeziums on same base and between same parallels as parent square	51
32, 33	Parallelograms on same base and between same parallels as parent square have the same area (i.e., area parallelogram = base × height)	51
34, 35	Area triangle = base × height/2	52

VARIOUS QUADRILATERALS WITH SIDES BISECTED

36	Rhombuses and rectangles: sides halved, areas quartered	52
37	Two families of similar parallelograms	52
38	Trapezium and parallelogram	53
39	Trapezium (symmetrical) and rhombus	53
40	Quadrilateral and parallelogram	53
41	Quadrilateral (symmetrical) and rectangle	53

VARIOUS QUADRILATERALS WITH ANGLES BISECTED

42	Square begets no figure, only a point	53
43	Rhombus begets no figure, only a point	53

44	Rectangle begets square	54
45	Parallelogram begets rectangle	54
46	Quadrilateral, parallelogram, rectangle, rhombus (sides and angles bisected alternately)	54

MISCELLANEOUS

47	Triangles, various quadrilaterals, and hexagon drawn on framework of fundamental circle drawing (Fig. 4)	56
48abc	Illustrating "angles in circle" properties on framework of fundamental circle drawing (Fig. 4)	56
49, 50	Demonstrating the theorem of Pythagoras	58-59
51abcde	Demonstrating the theorem of Pythagoras	59-60
52ab	Sketches of Crinoid (sea lily)	61

METAMORPHOSIS OF TRIANGLE, PENTAGON, AND CIRCLE (LEAF FORMS)

53a	Metamorphosis of triangle through three stages	62
53bcde	Metamorphosis of triangle through five stages	63
54	Metamorphosis of pentagon and construction drawing	64
55	Metamorphosis of circle and construction drawing	64

LINEWISE DRAWING OF CURVES

56, 57	Parabolas constructed linewise	65-66
58	"Curve of pursuit"	66
59	Logarithmic spiral constructed linewise	67
60-63	Curve designs constructed linewise	68

PENTAGON AND PENTAGRAM, GOLDEN RATIO

64	Pentagon and pentagram	72
65	Pentagram band or ribbon	72
66	Pentagon-pentagram construction in circle	82

67	Decagram band or ribbon	83
68, 69	Designs based on decagram and pentagram	84
70	Pentagon-pentagram figure	86
71	Angles of pentagon-pentagram triangles	86
72	Line divided in golden ratio	88
73	Construction for dividing a line in golden ratio	90
74	Family of golden rectangles	91
75	Comparison of angles of triangles drawn inside a golden rectangle	92
76	Golden compasses	92
77	American axe showing golden ratio proportions	95
78	Violin showing golden ratio proportions	95
79	Violin body showing detailed golden ratio proportions	96
80	Vase showing golden ratio proportions	97
81	Scale drawing, plan of seat of Coronation Chair, Westminster Abbey	98
	(NOTE: Figs. 79-81 are reproduced by kind permission of the Ministry of Works, London, Crown Copyright Reserved. Lettering showing golden ratio proportions has been added by the author.)	
82	Scale drawing, back elevation of Coronation Chair, Westminster Abbey	99
83	Scale drawing, side elevation of Coronation Chair, Westminster Abbey	100
84	Logarithmic spiral with radii vectors in golden ratio	104

ELLIPSE, HYPERBOLA, CURVES OF CASSINI, AND CIRCLES OF APOLLONIUS IN RELATION TO THE FOUR RULES OF ARITHMETIC

85abc	Ellipses (addition)	112-113
86abc	Hyperbols (subtraction)	114
87a-f	Curves of Cassini (multiplication)	116-118
88	Curves of Cassini shown in one figure	120
89	Curves of Cassini showing their orthogonal trajectories	120
90ab	Circles of Apollonius (division)	122
91	Circles of Apollonius shown in one figure	122

THE FIVE REGULAR SOLIDS

92	(a) Cube, (b) octahedron, (c) tetrahedron, (d) icosahedron, (e) dodecahedron	126

93	"Nets" for constructing the five regular solids	127
94	Plan views of five regular solids inscribed in circles of same radius	129
95	Octahedron and sphere inscribed in a cube	130
96	Tetrahedron contained by cube	131
97	Octahedron contained by a tetrahedron	131
98	Orthogonal projection of all five regular solids in one figure	132
99	Orthogonal projection showing reciprocal transformation of an icosahedron into a dodecahedron by producing the edges	132
100	Orthogonal projection of an icosahedron and dodecahedron inscribed in a cube	135
101a-d	Skeletons of various radiolarians, after Haeckel	136
102	Plan view of the small stellated dodecahedron	137
103	Plan view of the great stellated dodecahedron	137

THE CONIC SECTION CURVES: ELLIPSE, PARABOLA, AND HYPERBOLA

104	Elevation of cone showing cutting planes	143
105ab	Model of cone cut to give ellipse	145-146
106ab	Model of cone cut to give parabola	147-148
107ab	Model of cone cut to give hyperbola (one "wing")	149-150
108	Axis of symmetry between two points: a straight line	151
109	Axis of symmetry between a point and a line: a parabola	151
110	Axis of symmetry between a point and a circle enclosing it: an ellipse	151
111	Axis of symmetry between a point and a circle outside it: a hyperbola	152
112	Family of confocal parabolas	155
113	Family of parabolas with focus moving rhythmically	155
114	One family of confocal parabolas forming the orthogonal trajectories of another family	156
115	Family of confocal ellipses	157
116	Family of ellipses with one focus moving rhythmically	157
117	Family of similar confocal ellipses	158
118	Family of ellipses with one focus moving along the arc of a circle	158
119	Family of confocal hyperbolas	159
120	Family of hyperbolas with one focus moving rhythmically	159
121	Linewise construction of ellipse	160
122	Linewise construction of parabola	161

123	Linewise construction of hyperbola	162
124-126	Parabola, ellipse and hyperbola: comparing distances of a point on the curve from the focus and the directrix	164

PARALLELISM

127	Diagram illustrating concept of parallelism in relation to infinity	174
128a-h	Series of diagrams showing relative position of an object and its image formed by reflection in a concave mirror, illustrating the concept that there is one point at infinity on a straight line	174-175

CONTINUING THE CONIC SECTION CURVES

129-131	Ellipse, parabola, and hyperbola: illustrating that they may all be considered as curves of addition	176
132	Four identical parabolas	177
133	Two identical rectangular hyperbolas	177
134	The reflection of the sun's light and heat in a parabolic mirror	178
135	The path of a body projected at an angle of 45 degrees (a parabola)	178
136	Reflection of light from a spherical surface, spherical aberration, the caustic curve	179

PROJECTIVE GEOMETRY

137, 138	Hexagon inscribed in circle - Pascal's line	188
139, 140	Hexagon circumscribed round circle - Brianchon's point	189-190
141, 142	Pascal's line and Brianchon's point in relation to an ellipse	190-191
143, 144	Pascal's line and Brianchon's point in relation to a parabola	191-192
145, 146	Pascal's line and Brianchon's point in relation to a hyperbola	192-193
147abcd	Four examples of Pascal's line in relation to a circle	194
148	An example of Brianchon's point outside a circle	195
149	Partitioning of space: equilateral triangles, squares, hexagons	196
150ab	Pascal's line and Brianchon's point for pentagon as degenerate hexagon	196
151, 152	Quadrangle inscribed in circle - pole and polar	197
153, 154	Simple construction of pole and polar with reference to a circle (given the pole to find the polar; given the polar to find the pole)	198

155	Pole and polar and tangents to the circle	198
156	Quadrangle in circle treated as degenerate hexagon - Pascal's line is polar	202
157	Quadrilateral circumscribing circle and quadrangle inscribed showing self-dual nature of theorem of pole and polar	202
158,159	Correspondence between point inside triangle and line outside - pole and polar - particular case of theorem of Desargues	204
160-162	Illustrations of theorem of Desargues	205-206
163	Illustration for proof of theorem of Desargues - three dimensions	207
164	Part of Fig. 165	207
165	Desargues' configuration in one plane	209
166	Illustration of model in three dimensions showing proof of Desargues' theorem	209
167	Ten different cases of Desargues' theorem in one configuration	211-212
168	Harmonic quadrangles and harmonic points (harmonic conjugate pairs) ..	213
169	Family of harmonic quadrangles	213
170	Family of harmonic quadrangles	214
171	Harmonic conjugate pairs with one point at infinity	215
172	Family of harmonic quadrangles - perspective drawing of a chessboard ...	215
173, 173a	Relation between harmonic points and pole and polar construction with respect to a circle	216
174	Harmonic quadrangle construction in relation to Desargues' theorem	218
175	Harmonic points in a line (principle of duality)	218
176	Harmonic lines in a point (principle of duality)	219
177	Harmonic range in which ratio is 2:1	219
178	Dual nature of harmonic points in a line and harmonic lines in a point illustrated in one figure	221
179	Converse of Fig. 178 - projection of harmonic property	221
180	Harmonic conjugate points and the measure of a right angle	222
180ab	Reflection in spherical mirrors in relation to harmonic conjugate points ...	222
181	Fig. 180 in relation to circle of Apollonius	223
182	Construction of harmonic lines and points giving rise to harmonic quadrangles whose four points lie on a conic - in this case an ellipse	223
183	As for Fig. 182, except that the four points lie on a hyperbola	225
184	As for Fig. 182, except that the four points lie on a parabola	226
185	Method of constructing a conic from harmonic points and the Apollonian circle	227

186	Linewise construction of ellipse using method of Fig. 185	228
187	Linewise construction of hyperbola using method of Fig. 185	228
188	Linewise construction of parabola using method of Fig. 185	229
189, 189ab	Construction of harmonic quadrangle and diagonal triangle from a given conic	231-233
190	Three projections of a harmonic range	234
191	Method of constructing a conic using five random points and applying a sequence of projections	235
192-194	Pointwise construction of conic using method of Fig. 191	
	Ellipse	235
	Parabola	237
	Hyperbola	238
195	Drawing leading to a modified construction for a conic using five random points and five lines determined by these points	239
196	Construction of ellipse by method of Figu. 195 (five points and five tangents at these points) - partly pointwise and partly linewise	240
197	Modified construction of ellipse by method of Fig. 195 enabling a larger conic to be drawn in a smaller space	241
198	Projective construction of conic using five random points	241
199-201	Linewise construction of conic using method of Fig. 198 - parabola, ellipse, hyperbola	243-244

PLATES

1	Photograph of Coronation Chair in Westminster Abbey	101
2	Photograph of Coronation Chair in Westminster Abbey (back view)	102
	(Reproduced by kind permission of the Ministry of Works, London -Crown Copyright Reserved)	
3	*The Last Supper* by Leonardo da Vinci	105
4	*St. Francis Preaching to the Birds* by Giotto	106
5	*The Holy Trinity* by Masaccio	183
6	*St. George* by Donatello	184
7	*St. Jerome in His Study* by Albrecht Dürer	185

TABLES

6-1	Angular points, edges, and faces of the five regular solids	129
6-2	Metrical properties of the five regular solids	138
6-3	Metrical properties of the five regular solids	139

Editor's Comments

Geometry weaves its way as a central subject throughout the twelve-year curriculum of Waldorf schools. Through this subject, the teacher is able to explore both outer worldly realities as well as the inner world of humankind. It helps the children attain spatial harmony and develop their growing powers of analytical thinking.

In the first three grades, the children work at form drawing and dynamic drawing. In the fourth and fifth grades, they draw geometrical shapes freehand. In the sixth grade, they begin to use instruments and are challenged to draw figures with greater precision and exactness. At the same time, in eurythmy they execute geometrical exercises using their entire body. Also in eurythmy, copper rod exercises help secure the child's spatial orientation. In the seventh grade, two-dimensional drawing is continued, and the class spends a good amount of time understanding the theorem of Pythagoras, as well as perspective drawing and the areas of squares and triangles. In the eighth grade, exact constructions are continued with a focus on three-dimensional figures, the volumes of solids, the Platonic solids, and the laws of loci.

In the Waldorf high school, the students investigate descriptive geometry in the ninth grade— (Euclidean, coordinate, and solid geometry) as well as surveying in the tenth grade, projective geometry in the eleventh, and descriptive geometry as applied to practical problems in architecture in the twelfth. The quality of logical thinking that is exercised in these subjects and the students' attention to beauty and accuracy are considered very important.

This book by A. Renwick Sheen represents his life-long work as a Waldorf school teacher. It may prove to be invaluable to those of us working with Waldorf education and may be a stimulation to teachers from all philosophies of education. This text was first printed as a rough manuscript by the Waldorf Institute at Garden City in 1970. It is here completely re-edited and published by the AWSNA Publications through the courtesy of Stephen Sheen, his son.

— David Mitchell
Boulder, CO
2002

Preface

During the past 30 years, I have given many lectures in various European countries as well as in South Africa on the teaching of geometry. Frequently, after a lecture, I have been asked: "Is there anything published about what you have told us?" To date, I have had to reply that there are only a few articles in various journals and magazines. The present book is, therefore, an attempt to bring together much of this scattered information. At the same time, it is an expression of my teaching experience during this period to children of all ages at Michael Hall, the oldest Rudolf Steiner school in England. I must emphasize that nearly all that is in this book represents work that I have done at various times with different groups of children of ages from 3 to 18. Another teacher could well choose quite different themes and examples that would be equally suitable for a particular age group.

This book is written for those teachers of mathematics who are not content with the merely formal and logical presentation of their subject but who wish, as well, to present geometry as a cultural medium integrated with other subjects in the curriculum and who also recognize that children should experience the facts and laws of geometrical form before the logical proof is presented to them. Experience through drawing is of far greater value and importance to a child than logical proof without real inner experience of a truth. And, of course, such an approach can appeal to *all* children, whereas the logical proof treatment is often "over the heads" of a number of children in the class. I must emphasize that because little or no reference is made in this book to the more orthodox and formal aspects of geometry, this does not mean that they are omitted from the curriculum. They can be found in all the usual textbooks. For example, here will be found few formal proofs; proofs should certainly be given, but only at a later age after the children have gained the experience of a truth or law by accurate drawings. Or again, the analytical treatment of the conic section curves is an essential mathematical study for older children, but it should be preceded by such a treatment as outlined here to give a full understanding of the nature and properties of these important curves.

The method of presentation, then, is concerned with experience through drawing, leading to a stimulation of the children's imagination, which as de Morgan says, is "the moving power of mathematical invention." I have given careful descriptions of the various drawings and constructions, and I would warmly recommend that teachers do the drawings themselves before giving them to the children. In the wording of these descriptions, I have often deliberately avoided accepted mathematical terminology and used a more descriptive language, as I would do in teaching children.

It may well be raised that the time available for the teaching of geometry in a school curriculum simply does not allow for such a course of study as suggested in this book, especially when one takes into consideration the more formal aspects of geometry, which have been omitted. This is a quite valid objection. I must, however, point out that the course of work described here is not meant to be given in its entirety. The teacher will naturally choose what he or she feels most suitable for a particular group of children. It may also be mentioned that in a Rudolf Steiner school, there is a considerable economy of teaching time, in that what are called Main Lesson subjects—of which geometry is one—are taught in periods of three to four weeks for two hours each day (first thing in the morning). This, of course, means that the children are able to concentrate on a subject, and they can really get a great deal done during such a period.

I would like to thank my colleagues, Mr. H. Gebert, B.Sc., Mr. John Davy, B.A., and Bengt Ulin of Uppsala, for their valuable advice and helpful criticism. I am also indebted to Mr. George Adams, M.A., for essential help, especially in connection with the chapter on projective geometry. To Herr Ernst Bindel, mathematics teacher in the Waldorf School, Stuttgart, and to Dr. Hermann von Baravalle, former mathematics teacher in the same school, I wish to tender my thanks for their kind permission to refer to work that they have done and that I have often used in my own teaching. My grateful acknowledgments are also due to the several publishers and authorities mentioned in the text who have granted me permission to publish quotations, diagrams, and pictures.

Finally, this book could never have been written without the fundamental inspiration of the late Dr. Rudolf Steiner, the great teacher of teachers for our age.

— A. Renwick Sheen

Chapter 1

Introduction

Over 100 years ago, the German philosopher and educationist Herbart (1776–1841) said that he could not imagine any instruction that was not at the same time education, and vice versa, he could not imagine any education that could dispense with instruction. Today, the majority of teachers in schools of all kinds would fundamentally agree with such a conception. On the other hand, in actual practice in the classroom, a great deal of the teaching given to children of all ages is mere instruction and has little or no relation to their moral nature or to the development of character. These deeper aspects of education are too often left to what is sometimes vaguely called the "school atmosphere," though many boys and girls owe a great deal to the fine influence of one or other individual teacher who, through his or her personality, has brought real moral strength into their lives. The same problem is discussed by Lord Elton in his book *St. George and the Holy Grail* (1942), in which he insists that all imparting of knowledge should have a religious, moral quality. He says, "A Christian education is a particular kind of education in all subjects." He also points out that, at the time he is writing, there is one great country of Europe where there is certainly no divorce between education and instruction: "The Nazi school is not a school which devotes an hour a week to teaching a certain creed, but a school which teaches everything in a certain way." It is a strange thing that a fine ideal can be so degraded by a wrongful application, and it is one of the ironies of life that this can so easily happen. If teachers throughout the world today could and would bring the great human and moral values into the classroom through the actual material of their lessons with anything like the force and thoroughness with which the Nazi teachers inculcated inhuman and immoral values into the German children of that generation, then they should be well on the road towards achieving such an educational ideal as that outlined by Herbart. The reason for our failure to do this is very largely due to the great emphasis laid today on an intellectual form of teaching. The training of the child's intellect is considered of paramount importance; this is the result partly of the pressure of examinations at various ages but also of the modern view of child psychology, which tends to look upon children as little grown-ups, with the same kinds of faculties as the adult, though less developed. Every human being expresses himself in life through the three faculties of thinking, feeling, and willing. In the being of the child, these simply are not there in the same way as in the adult man or woman. For example, the thinking of a child of 10 or 11 years of age is intimately bound up with the life of feeling. It is a feeling-thinking and not yet a logical, intellectual thinking. The soul expression of children of this age is fundamentally in the realm of feeling and imagination, and education at this time must go with this and not try to call forth powers that are not yet properly awake.

In a letter to a friend in 1817, Keats wrote: "I am certain of nothing but of the holiness of the Heart's affections and the truth of Imagination. What the imagination seizes as Beauty must be Truth . . . The Imagination may be compared to Adam's dream—he awoke and found it Truth." This is the inner experience of a young child, and as a poet, Keats carried this wonderful power of the imagination into adult life. Today, our modern intellectual education provides little food for the imagination, and so this most precious gift of childhood tends to wither and die instead of growing and developing into one of the noblest expressions of the human soul. The really great people of all ages have been people of imagination. The great and noble deeds of history are the outcome of the vision of such people and have often had the most far-reaching effects on the lives of men and women the world over. Just to take one example from our own day: Sir Winston Churchill, combining imaginative vision with an indomitable will, more than any other single man, saved the countries of Europe from subjugation by a ruthless tyranny. It is worth noting that, according to his own confession, he was a dull scholar at Harrow, judging by ordinary intellectual standards. Perhaps it was just because he unconsciously warded off the deadening effects of over-intellectualism during this early school years that he was able to preserve into later life a great power of imagination and immense forces of will.

In the light of what has been said, we are now in a position to give answers to some very important questions that should surely be asked concerning every subject that a child learns at school. Why should children learn geometry? Assuming there are good reasons for its inclusion in the curriculum, then when should the first lessons in geometry begin? And lastly, how should geometry be introduced and continued through the school?

Why should children learn geometry?

There are two obvious answers: Everyone needs to know the elementary laws and properties of the simple geometrical figures—the circle, triangle, square, rectangle, and so on—which we see everywhere around us and which we use in all kinds of construction; then the study of these laws and their sequence of proof as in Euclid is a fine exercise and training for the powers of logical thinking. The first of these reasons is clearly of utilitarian value and belongs more to the sphere of instruction only; the second reason combines instruction with education. But there is still a much more fundamental reason for the inclusion of geometry in a school curriculum, and it is with this that this book is chiefly concerned. Plato once said, "God is eternally geometrizing," and if we can gain some idea of what these words really signify, we shall realize that geometry is a fundamental subject in all education and has not only the utilitarian value referred to above but can become a source of deep moral significance for the whole of life. Everywhere around us we see the manifold forms of nature: the great variety of crystal shapes with planar faces and sharp, straight-line edges; the infinite metamorphoses of form in the plant world, mostly curved forms; the complexity of shape of different animal species; the form of man; the various microcosmic forms of cell structure within the kingdoms of nature; the majestic macrocosmic curves traced out by the heavenly bodies in their orbits. These are all expressions of geometrical law at work in the earth, on the earth, and in the universe. Furthermore, if we can learn to know something of the real structure of space itself, then we shall achieve a more fundamental understanding of the manifold forms that exist within space. A study of projective geometry will help us towards such a knowledge, and the last chapter of this book will suggest how this subject may be presented to older children.

Concerning projective geometry, Morris Kine, professor of mathematics, New York University, says, "The contents of all four geometries are now incorporated in one harmonious whole!" This aspect of geometry is, therefore, of fundamental importance in this book, and its forms of thought have been introduced into earlier chapters. It is therefore evident that geometry is an essential subject of a school curriculum from the point of view both of instruction and education, and we may now consider the next question.

When should the first lesson in geometry begin?

Just as there is the right moment for the little child to stand upright and take her first unsteady steps, so there is the right moment for her to begin to learn the three R's, to hear about the history of the Romans for the first time, to be introduced to the world of scientific phenomena in physics and chemistry. The modern tendency is to hurry everything on and to teach children subjects, or different aspects of subjects, too early, before their inner development is mature enough for them to really grasp what they are being taught. This often has the effect of forcing the intellectual development prematurely, and this, of course, can be done, but only at the expense of other faculties, which thereby become enfeebled—for example, the imagination. No sane mother would ever encourage her small child to walk before she saw that he was ready for this great adventure; she knows that if she did, the result would be disastrous and probably lead to deformity for life. In the sphere of the soul, the effect of bringing intellectual teaching too early to children is crippling to their full human capabilities and may even lead to actual bodily infirmities in later years. Many parents today are proud of the fact that their boy or girl can read, write, and do simple sums at the age of four or five. How much better it would be for the future development of their child if they were proud of the fact that he could not do any of these things at this age in spite of the efforts of his teachers!

This modern tendency is illustrated most strikingly by an advertisement that appeared about thirty years ago in one of the leading English Sunday newspapers. It was headed: "WANTED—A SCIENTIST of the first order, if necessary of senior standing, but as young as possible, with a knowledge of the theory of science, to investigate and conduct the introduction of young children, 4–10, to science and scientific method." Then followed a long statement of the problem, which contained this sentence: "It is as yet uncertain whether there exist any special factors limiting or making undesirable the introduction of children of 4–10 to scientific knowledge and scientific thought." Towards the end, the advertisement said: "In order to be able to obtain the services of the man most suited to the work, they are advertising widely and they are prepared to pay such salary as will enable him to leave his present occupation, whatever that may be." The word *they* refers to the directors of the school who inserted the advertisement, which ends with the names of two very eminent scientists and one very eminent educationist who "have kindly consented to assist the directors in the final selection of candidates." Perhaps a fair comment would be that parents and teachers who have any real feeling for the being and nature of a little child should be as indignant and angry at such a project as they are when, from time to time, they read in the daily press of a case of physical cruelty suffered by a child, and they might well ask for the establishment of a National Society for the Prevention of Mis-education of Children. To be just to the promoters of the scheme, they obviously considered it in the nature of an experiment involving, by their own confession, possible uncertain factors that might make it undesirable to continue the experiment. Such is the ignorance of modern humanity concerning their own nature and inner development, that these educationists failed to recognize that the one quite certain factor before them was that scientific knowledge and especially scientific method, as these are conceived today, have no relation whatever to the real natural inner development of children from 4 to 10 years of age.

In 1935 a noteworthy book was published with the challenging title of *Man the Unknown*. The author, Alexis Carrel, was the chief biologist at the Rockefeller Institute in the United States, and the theme of his book was that the great scientific achievements of our age do not include any real knowledge of the nature of the human being. Although this book was written twenty years ago and the advertisement appeared some ten years earlier, science has made little or no advance towards an answer to Carrel's challenge, and educationists still carry out experiments on children, which through ignorance of their "material," may often prove highly dangerous. The headmaster of a well-known boys' school once said that he considered modern education to be a criminal occupation, and undoubtedly he regarded himself as one of the arch-criminals!

And yet Carrel's challenge has been answered but not from the direction of orthodox science, and moreover it was answered in the early years of this century long before his book appeared. In Rudolf Steiner's teaching, modern man can come to a real understanding and knowledge of his true nature in body, soul, and spirit. If we are not willing to study such a knowledge, it is simply because we are at the mercy of our ordinary habits of intellectual thinking and will not make the effort to break through to new conceptions. In his book, Carrel says: "We cannot undertake the restoration of ourselves and our environment before having transformed our habits of thought." The philosophy of Rudolf Steiner would fully endorse this statement, and furthermore he shows us how we may carry out the transformation.

Today, education need no longer be "a criminal occupation," for in 1919 with the founding of the famous Waldorf School in Stuttgart there came into being an educational movement that has grown to include over 900 Rudolf Steiner Schools around the world. Teachers in these schools do not experiment with children but strive to educate them through insight into and knowledge of their natural inner development derived from the "picture of the human-being" as given by Rudolf Steiner. (For those readers who may wish to know more about the principles and ideals of Rudolf Steiner education, a selected list of books and pamphlets will be found at the end of this book.)

To return, then, to our question, we must first consider the essential nature of our subject, geometry. Among all subjects in a school curriculum, this is one that appeals fundamentally to the intellect and demands for its study a thinking that is clear and logical. Certain aspects of geometry, as we shall see, also require for their understanding and interpretation a thinking that is not only logical but has a certain imaginative quality—an imaginative thinking. Now the child of elementary school age up to about 12 years lives mainly in the experience of feeling-imagination. The words of Keats quoted earlier belong essentially to this age of childhood. The thinking of such a child is a feeling-thinking. The logical nature of the life of thought does not naturally come to full expression in the soul of the child until the time of puberty at about the age of 14 years. (Of course, as we have pointed out, logical thinking can be and generally is "forced" prematurely in much of our mis-education today.) The process, within the soul of the child, of passing over from a mere feeling-thinking to a logical thinking is a process in time, and we find that in most normal children it starts at about the age of 12 years. In other words, we may say that at the age of 12 the boy or girl begins to anticipate the change that is coming at puberty when the faculty of logical thinking becomes awake. So if we bring together the essential nature of the subject to be taught and the inner soul quality of the child, we see that the teaching of geometry should begin at about the age of 12 years. At this moment, provided the subject is introduced in the right way (see Chapter 3), children have a real inner relationship to geometry.

How should geometry be introduced and continued through the school?

It is the object of the following chapters of this book to give one answer to this question.

The author is aware that, during, say, the past eighty years, attempts have been made by individual teachers in many different schools to bring a more imaginative element into the teaching of geometry. One such noteworthy contribution is that of Mrs. Edith L. Somervell, who wrote a valuable little book entitled *A Rhythmic Approach to Mathematics*[1] as long ago as 1906; her ideas were put into practice in several schools including Bedales. (Reference will be made again to this work in Chapter 3, "First Lessons in Geometry.") There have been many children who have greatly benefited from such imaginative methods used by individual teachers who have had an instinctive insight into the real needs of their pupils. At the same time, it must be recognized that the usual approach has been, and still is, the intellectual one demanded by the examination system. In what follows, the attempt has been made to indicate a method of treatment of the subject that springs not just from the ideas of an individual teacher but belongs essentially to the ideas and ideals of Waldorf education as a whole.

Chapter 2

Pre-Geometry

In the previous chapter, we have seen that the right age at which children should begin lessons in geometry is about 12 years. Before this age, however, in many schools today, children gain an experience of geometrical forms in other lessons, and this is an excellent preparation for the more formal work they will do when they are older. One might even say that it is an essential preparation.

Most teachers today recognize that, especially for little children, movement and rhythm are a very important means of expression and must take their place in a well-balanced curriculum. In some schools, lessons are given in Dalcrose eurhythmics, while in others, teachers will "invent" their own movement lessons. In a Waldorf school all children are taught eurythmy, an art of movement which may be described as visible speech and visible song (see List of Selected Books). In such lessons, the children move in certain forms on the floor of the room—circles, squares, triangles, pentagrams, lemniscates (figures of eight)—and they should learn to walk or run those forms accurately and precisely. So they experience many of the forms of geometry through the body in rhythmic movement, and such activity can give children a very fundamental feeling for form. Actually, all our experience of outer form comes to us through movement, though generally we are not conscious that this is so. We may go into a great cathedral and, standing in the nave, observe the form of a Gothic arch carried by the pillars on either side of us and in such an observation become aware that our gaze is travelling along the stone tracery of the arch. So in every observation of form, however small, our eye "travels round" the form, but generally the movement is so slight and so quick that we are not conscious of it. In fact, we are really only conscious of such a process when the following of a form with the eye involves moving the head. Thus, movement and the experience of form are closely related, and little children who have an opportunity to carry out such exercises in movement will gain a firm inner feeling for the forms they will later study in their geometry lessons.

Another activity that small children enjoy is free-hand pattern drawing, and if directed to express principles of balance and symmetry, this too is a good foundation for later work. From quite an early age, a child experiences the symmetry of right and left, which belongs to his own body, and this inward experience he is quite ready to express outwardly in the drawing of balanced, symmetrical forms. The following drawings (Fig. 1) are copies of such work done by children of between 7 and 9 years; the actual drawings were, of course, carried out with colored crayons.

In early arithmetic lessons, the relation between number and form should be introduced. For example one may arrange 12 golden stars (cut out of paper) evenly spaced round a circle against a dark blue background. Now ask the children to take stars away so as to leave first 6 and then only 3 in as good an arrangement as possible (Fig. 1a). It is interesting to observe how some children will immediately see what to do so as to arrive at the hexagon form of the number 6 or the triangle form of the number 3, while others will remove the stars so as to leave quite unsymmetrical arrangements. Or again, one may ask the children to take 4 or 8 stars and arrange them in a perfect form or pattern (Fig. 1b). It is good even at this early age (7 or 8 years) that children should come to realize that each number is related to a form or pattern. They naturally enjoy seeing how the 6-form (hexagon) arises from the 12 by taking away alternate stars and how again by removing alternate stars, the 3-form (triangle) is left.

Fig. 1

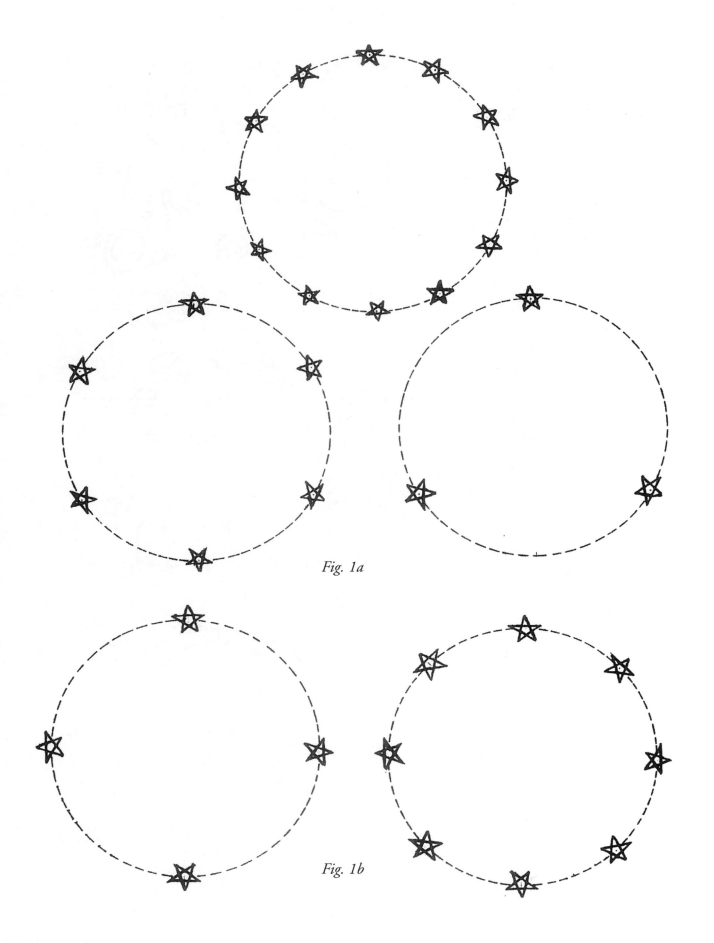

Fig. 1a

Fig. 1b

A very vital principle of all education is that of metamorphosis. Something that is taught to a small child may in later years become transformed into a capacity. It is a well-known fact today that the slumbering intellectual powers of many backward children can be awakened into life by working through the activity of the limbs, for example, in various kinds of handwork. All children of 7 or 8 should be taught to knit, not only because knitting is a useful thing to learn, but because the activity of joining one stitch to the next with great exactness can become transformed in later life into a capacity for clear logical thinking. We even speak of "well-knit thoughts." Respect for authority during childhood is the foundation for self-confidence and inner responsibility in the adult. Or again, if we foster the natural reverence and wonder that the small child has for all that is around her—and how little this is done today!—then such feelings are metamorphosed into moral strength and the power to give blessing in old age. So, if during the school years, children are given a strong feeling for balance and symmetry of form, then when they are grown up they are more likely to be able to make balanced and wise judgments in life. If we become more and more sensitive to these relationships, we shall realize what an immense amount we can do in the classroom, through the actual material of our lessons, towards building the character and inner moral strength of a future generation.

It is customary today to give children intelligence tests as a means of determining the group to which they belong so that children with similar I.Q.s may be taught together. Surely an I.Q. tells us about only a part of a child's spiritual capacities, and as educators we should wish to know far more. The incapacity of a child to express herself in the drawing of symmetrical and balanced forms is a very important indication of her development, and tests carried out by such simple exercises—taken, of course, in connection with other observations—will often reveal to the educator the necessity for special treatment and special methods of education.

Chapter 3

First Lessons in Geometry

The first lessons in any new subject are of the greatest importance, and the teacher should take special care in preparing such lessons that they may make a deep and lasting impression on the receptive mind of the child. The wise teacher will tell his or her children that in a few weeks' time they will begin to learn a new subject—geometry. She may tell them that they will be doing accurate drawings with a very sharp pencil, using ruler and compass. To lead children in this way towards the "unknown" or "only half-known" is to arouse their interest in and eagerness for the lessons that are to come. The word g*eometry* will be a new and mysterious name to awaken in the children keen anticipation. If the lessons are to live up to this expectation, the teacher will have to answer this question: Within the sphere of geometry, what is there that belongs quite naturally to the ordinary experience of children of 12 years old? If such a question concerning this and other subjects were taken seriously, young children would be saved a great deal of abstract intellectual teaching that has no relationship to their own inner nature. This is especially the case in the more scientific subjects in the school curriculum. Professor J. J. Findlay in his book, *The School,* refers to this problem: "We may take as an example the efforts at reform made by Herbert Spencer (1861), exerted through those four famous essays on Education. Spencer voiced the progressive opinion of his time on behalf of science: The children are being deprived, he declared, of knowledge about all that lies in most direct relation to their needs; they possess bodies and are ignorant of physiology; even the mothers and daughters know nothing of the laws of nature which underlie the vital concerns of our domestic life. He succeeded in moving public opinion; those who controlled the schools became genuinely in earnest to arouse in the young a desire to learn about these things, and South Kensington examined millions of children in physiology and the laws of health. The result certainly has not met the hopes of those who started the scientific movement in schools. Undoubtedly they have succeeded in establishing the prestige of science: It is placed side by side with the more venerable pursuits of the academy, and all the machinery of learning is available on its behalf: It takes rank, with its logic, its text-books, its examinations, as part of the cultural system; the scientific man claims equal rank with those who profess more venerable cults. But, while thus elaborated and organized to suit the tastes of adult thinkers, the pursuit of science loses the very qualities that make it of service to the young. Chemical atoms have no advantage over Greek particles unless the advantage becomes part and parcel of values realized and appreciated by the school-boy; physiology may be immensely important for the welfare of mankind, but all the exhortations of anxious reformers will not compel the young to care about it unless it can be brought into relation with their crude and unorganized experience."

What then, within the range of the subject we are considering, belongs to the child's "crude and unorganized experience?" There is, of course, not just one answer to this question, and so the teacher is free to choose his starting point. What now follows is the answer of one teacher; others might choose a different approach.

Let us consider the form of the human being. Here we see a fundamental polarity of form expressed between the rounded spherical shape of the head and the straight line of the limbs. (We need not be concerned at this stage that the head is not an exact sphere, nor that the arms and legs do not fulfill the definition of a geometrical straight line.) The point is that the roundness of the head and the straightness of the limbs belong to the experience of every child, and this polarity of form is expressed everywhere in nature. Between roundness and straightness there lies every kind of curvature. Straight lines and plane (flat) surfaces belong fundamentally to the mineral world as we see in all the different crystal structures. In the world of life—plant, animal, and man—we find an endless variety of curved forms. In the movements of sun, moon, and stars, we follow majestic circlings across the heavens from their rising to their setting. So one directs the children's attention to the manifold forms in nature and natural processes. Next, one can pass over to a consideration of man-made forms—for example, in architecture, artistic creations, machinery, and so on: the simple roundness of an Eskimo igloo, the spherical dome of a great cathedral, the square tower and delicate triangular spire of a church, the rounded Roman and pointed Gothic arches, the rectangular forms in the classroom, the fine curve of a bridge spanning a river, the beautiful curves of a Greek vase, the complicated forms of a violin, the combination of roundness and straightness in a wheel with its spokes, and so on. The children will be eager to give these and many other examples of form in nature and in the world around them.

The circle has always been considered as the most perfect of geometrical figures. Right up to the time of Kepler (1571–1630), astronomers and philosophers conceived that the heavenly bodies, being themselves perfect, must move in circular orbits, and even he was reluctant to abandon the beauty and simplicity of circular motion in favor of the ellipse (which, after all, is only a projection of the circle). With such a perfect and fundamental figure, then, children may begin their lessons in geometry. They will delight in using their compasses and, by doing various drawings and designs, will learn the simpler laws of the circle. (The measurement of the circle and the significance of π do not concern these first lessons.) The first two drawings (Figs. 2 and 3) illustrate the important fact that circles are always exactly the same shape and that one circle can only differ from another in the matter of size. The children should observe that in drawing a circle, their compasses remain stretched always the same distance, and they must learn to hold them so that they do not squeeze them! This will lead them to see that the circumference is everywhere the same distance from the center and that by altering this distance (the radius) they can draw larger or smaller circles. The second drawing (Fig. 3) clearly demands much greater accuracy than the first. The third drawing (Fig. 4) may be considered as the fundamental circle design and again needs real accuracy in the use of the compass. It illustrates the fundamental law of the circle that the radius can be stepped off exactly six times round the circumference. Figs. 5 and 6 illustrate designs involving a repetition of Figs. 3 and 4, and Fig. 7 shows the circling of a circle. Of course, after such work has been carried out accurately in fine pencil line, the drawings should be artistically colored, shading with colored pencils. (This also applies to all the drawings that follow. Sometimes it is better only to color the lines and not do any colored shading. The coloring should also be related to the form and to the changing or movement of the form if this occurs.) This activity of coloring is not a waste of time; on the contrary, it gives the children a greater understanding for the form they have drawn and also an enhanced appreciation of its symmetry and beauty. Other and more complicated circle designs will be illustrated later involving straight-line forms.

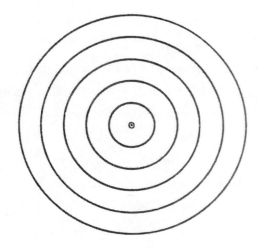

Fig. 2

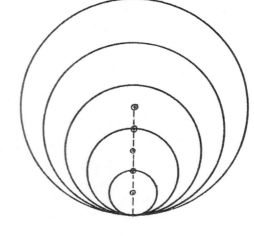

Fig. 3

Fig. 4

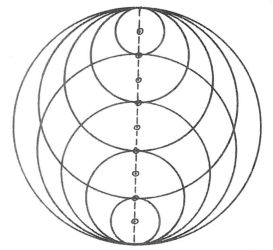

Fig. 5

Fig. 6

Fig. 7

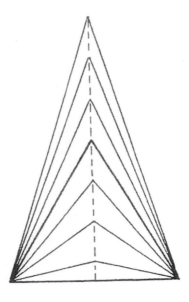

Fig. 8

The simplest of straight-line geometrical figures is the triangle, and our method of teaching geometry in pictures will now concern itself with this figure and its laws and properties. Unlike the circle, the triangle may have many different shapes, and we start with the artistic, perfect form, the equilateral triangle, and evolve the others from it as illustrated in the two following drawings (Figs. 8 and 9). In the first one, the equilateral triangle is metamorphosed into a family of isosceles triangles, and in the second, into a series of scalene triangles. The children should become quite conversant with these names and their meanings: equilateral (Latin), equal sides; isosceles (Greek), equal legs; scalene (Greek), limping, uneven. A man standing with his legs apart makes an isosceles triangle with the ground; if he limps as he walks, then one leg is longer than the other and his legs make a scalene triangle with the ground. Again the children can invent many designs on the basis of these two drawings, which, with careful coloring, afford them real pleasure. Such design is shown in the third drawing (Fig. 10).

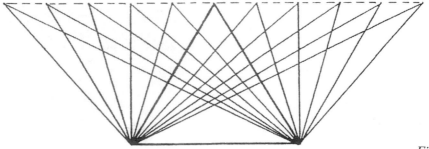

Fig. 9

Fig. 10

Before proceeding to the naming of triangles by their angles the concept must be made clear:

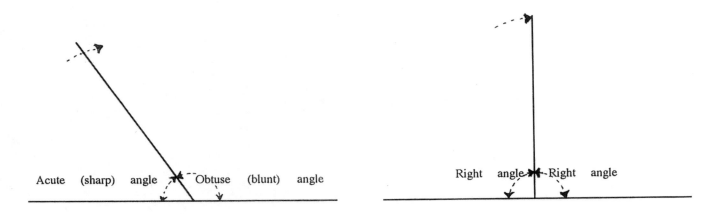

Fig. 10a

The right angle is formed by the balancing position of the rotating arm when the angle is neither acute nor obtuse (Fig. 10a). Thus we can have acute-angled triangles, obtuse-angled triangles, and right-angled triangles. To ensure that the children are quite familiar with the different kinds of triangles, it is a good exercise to ask them such questions as: Is it possible to have a right-angled equilateral triangle or a right-angled isosceles triangle? Is an obtuse-angled isosceles triangle possible? From the drawings they have done, they will easily see that an equilateral triangle has not only all its sides equal but also all its angles equal, that an isosceles triangle has two equal angles, and that in a scalene triangle the angles are of different sizes. Now there will follow a series of drawings illustrating the laws and properties of triangles, and these will involve such simple constructions as bisecting a line, bisecting an angle, drawing one line at right angles to another from a point in the line or from a point outside it, and the drawing of one line parallel to another.

Fig. 11   Family of equilateral triangles; sides are halved and areas are quartered.

Fig. 12   Family of right-angled triangles; angle in a semi-circle is a right angle.

Fig. 13   Exterior angle of a triangle equals the sum of the two interior opposite angles. Also, alternate angles between parallel lines and a transversal are equal, and corresponding angles between parallel lines and a transversal are equal.

Fig. 14   The three angles of a triangle are together equal to two right angles.

Fig. 15   Equilateral triangle divided into six right-angled triangles. Fold about each of middle lines in turn and a shaded triangle comes on a white triangle. Example of three-fold symmetry.

Fig. 16   Movement of equilateral triangle towards and beyond the center of a circle.

Fig. 17   Interlaced equilateral triangles in a circle.

Fig. 18   A modification of Fig. 17.

Fig. 19   A 24-sided regular polygon in a circle with all its diagonals. Each of the 24 angular points is joined to every other one, and there are altogether 276 straight lines in the figure.

This last drawing (Fig. 19) is one that many children greatly enjoy doing, and it is a good exercise for joining two points accurately by a straight line. It is fascinating to see how the lines weave together to form the envelopes of concentric circles expanding rhythmically.

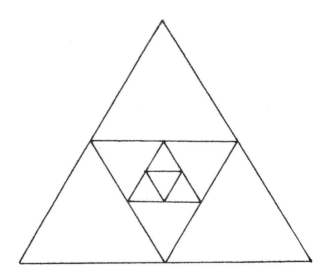

Fig. 11

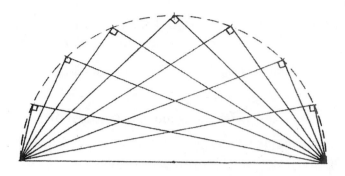

Fig. 12

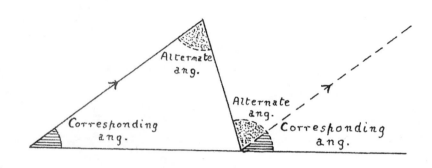

Fig. 13

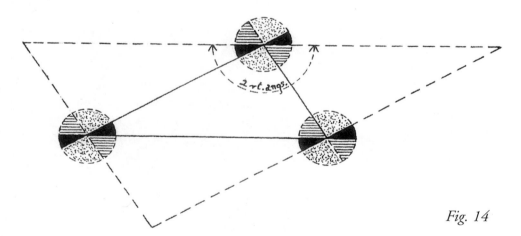

Fig. 14

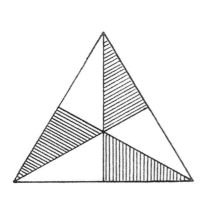

Fig. 15

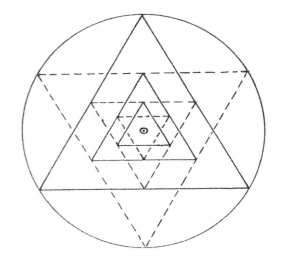

Fig. 16

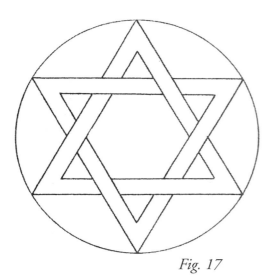

Fig. 17

Fig. 18

Fig. 19

We now come to four-sided figures—quadrilaterals—and take our start from the most perfect, that is, the square, which has all its sides equal and all its angles right angles. The following drawings are designs based on the square. Fig. 20 corresponds to the triangle drawing (Fig. 11); here again, the sides are halved and the areas are quartered. Figs. 21 and 24 are star designs in which we see the figure of the octagon arising and also that the line forming the star is "continuous," returning to its starting point. Again, these designs involve constant practice in the simple geometrical constructions.

From the square we pass over to the next most perfect quadrilaterals: the rectangle, which has all its angles equal (right angles) and its opposite sides equal, and the rhombus which has all its sides equal and its opposite angles equal. Figs. 26 and 27 show families of these figures arising from a square.

Fig. 26 is of particular interest and importance in that the rectangles and the square have been so drawn that they are all of equal area. It will be noted that if the corners of the rectangles and square are joined by smooth curves, we obtain two rectangular hyperbolae. If now we draw in a similar configuration a family of rhombuses having the same area (Fig. 27) (the angular points of such rhombuses move along the rectangular axes in geometrical progression), then we see that the sides of the figures form a pair of rectangular hyperbolae. That is, the curves may be drawn touching the sides. If sufficient rhombuses are drawn, then the hyperbolae appear just as the circles do in Fig. 19. Thus in Fig. 26, we have a foundation construction for a point-wise hyperbola and in Fig. 27 for a line-wise hyperbola. Of course at this age we are not concerned with considering this curve, but it is of importance that children should carry out such elementary drawings that have within them real significance for more advanced work.

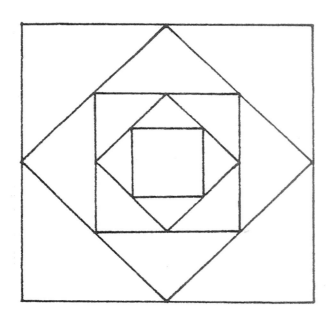

Fig. 20

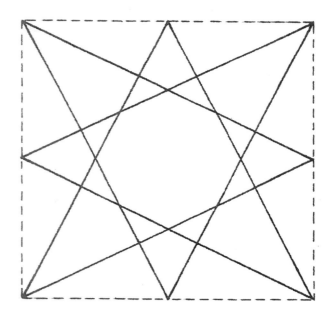

Fig. 21

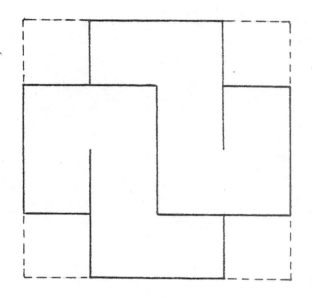

Fig. 22

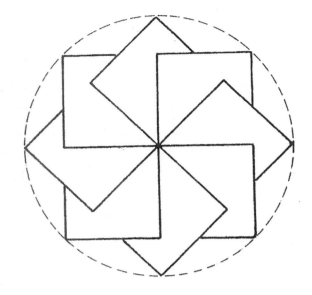

Fig. 23

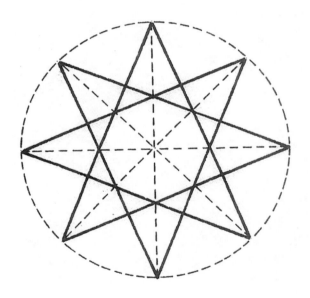

Fig. 24

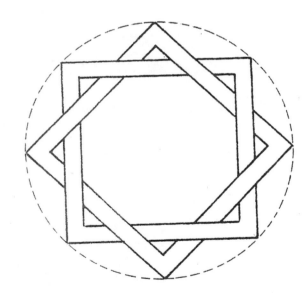

Fig. 25

46

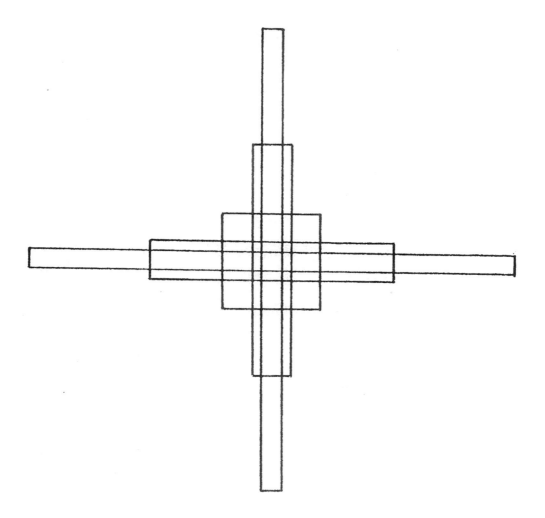

Fig. 26

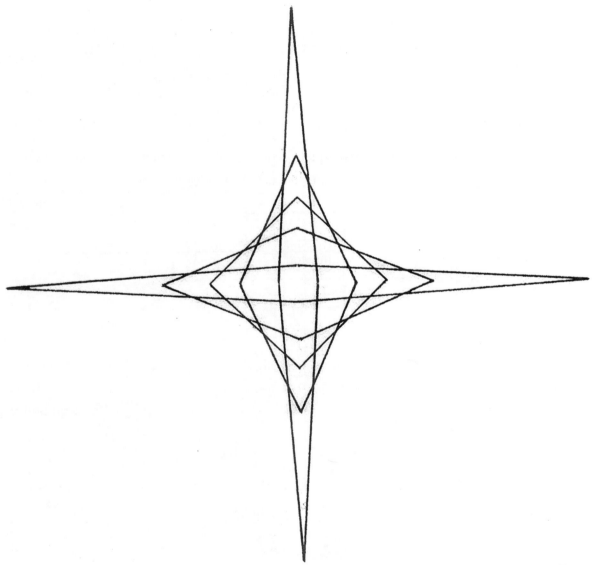

Fig. 27

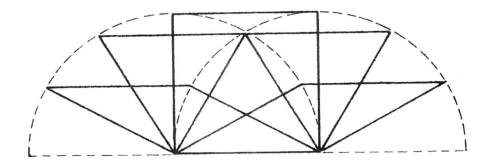

Fig. 28

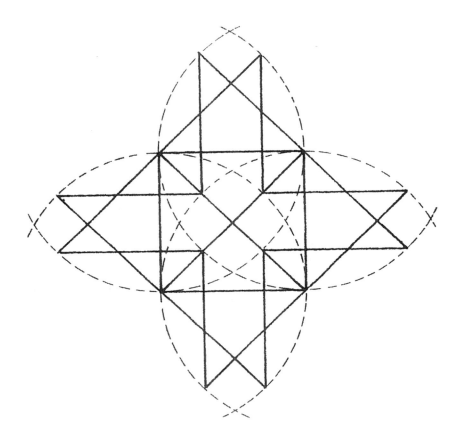

Fig. 29

When the children come to study the conic section curves at the age of 15 or 16, then the teacher will be able to refer to such earlier work. Fig. 29 is a design of rhombuses, two on each side of the square; it is an interesting figure and, although built up out of eight rhombuses, derived from a square. Eight more squares appear of two different sizes and many right-angled isosceles triangles of three different sizes.

Following these more regular quadrilaterals, we come to the parallelogram, which has opposite sides parallel and opposite angles equal, and the trapezoid, which has only one pair of parallel sides.

Figs. 30 and 31 show families of parallelograms and trapezoids developed from a square. The remaining possibilities are irregular figures that are simple, called quadrilaterals.

In Fig. 30 we have a family of parallelograms on the same base and between the same parallels as the "parent" square, or, expressed in another way, the parallelograms and the square are on the same base and have the same height. Figs. 32 and 33 demonstrate the fact that the parallelograms all have the same area as the square, that is, the area of a parallelogram is the product of the base and the height. From this, it follows directly that the area of a triangle is half the product of the base and the height, for every triangle can be considered as half a rectangle or parallelogram (the diagonal of a parallelogram bisects it) (Figs. 34 and 35).

In the next set of drawings, we see how one or other four-sided figures can arise from another quadrilateral by a simple construction, that is, the bisecting of the sides or the angles. Fig. 20 shows how a square arises when we bisect the sides of a square; here, we may say, the perfect form "gives birth" to the perfect form. The next two most perfect forms, the rhombus (equal sides) and the rectangle (equal angles), produce one another alternately by the same construction—bisection of sides (Fig. 36). This drawing is of real significance for later studies when the children are much older and learn some of the fundamentals of projective geometry (see Fig. 1). For the families of rhombuses and rectangles arising in this way, we see that the sides are halved and the areas are quartered. By halving the side of a parallelogram we get another parallelogram (Fig. 37); two families of similar parallelograms arise. By the same construction, a trapezoid creates a parallelogram (if the trapezoid is symmetrical, the parallelogram is a rhombus), and a quadrilateral also creates a parallelogram (if the quadrilateral is symmetrical, the parallelogram is a rectangle) (Figs. 38, 39, 40, and 41).

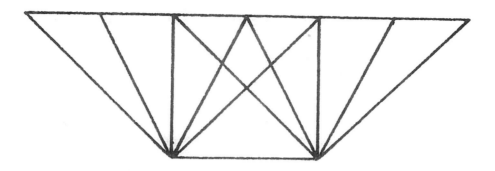

Fig. 30

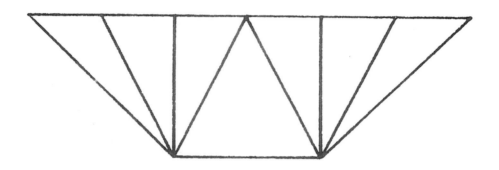

Fig. 31

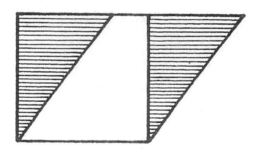

Fig. 32

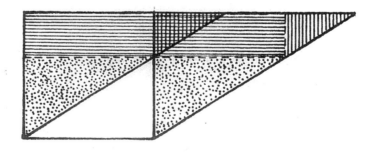

Fig. 33

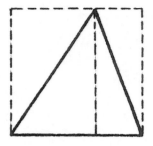

Fig. 34

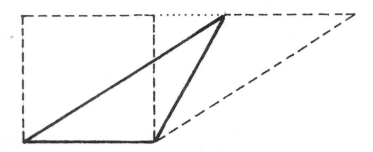

Fig. 35

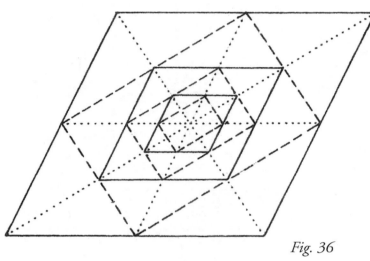

Fig. 36

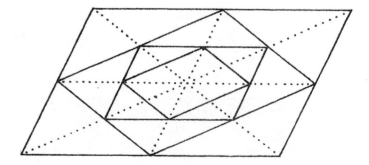

Fig. 37

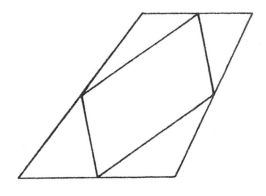

Fig. 38

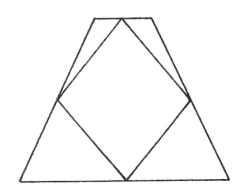

Fig. 39

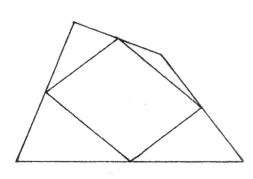

Fig. 40

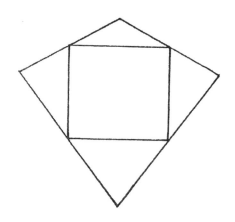

Fig. 41

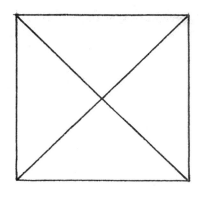

Fig. 42

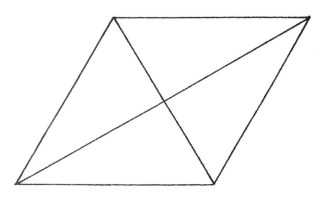

Fig. 43

53

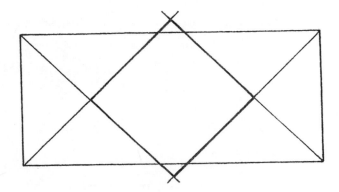

Fig. 44

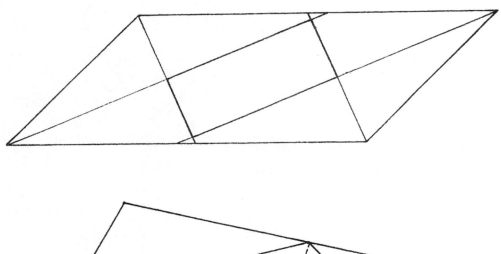

Fig. 45

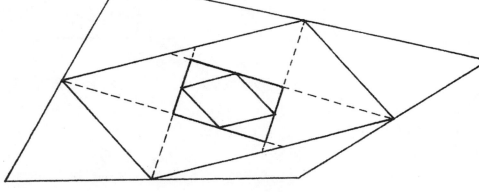

Fig. 46

If the angles of a square or of a rhombus are bisected, no figure arises; we get only a point (Figs. 42 and 43). In both these figures, the diagonals bisect the angles. A square is formed when we bisect the angles of a rectangle (Fig. 44), and a rectangle arises when we bisect the angles of a parallelogram (Fig. 45). Fig. 46 is a combination of previous drawings, starting with a quadrilateral and alternately bisecting sides and angles of the resulting figures.

Again considering this principle of bisecting the angles of a quadrilateral, it is of interest to approach the problem from the parallelogram (Fig. 45). Since the opposite angles are equal, it is clear that the bisectors of these angles are two pairs of parallel lines that create a rectangular cell; as the sides of the parallelogram become more and more nearly equal, these parallel bisectors move closer and closer together and the rectangle becomes progressively smaller. When the parallelogram becomes a rhombus, that is, when the sides are all equal, the rectangle shrinks to a point and the parallel lines coincide to form the diagonals of the rhombus (Fig. 43). The four right angles of the rectangle remain but are transformed into the four right angles of the crossing diagonals. The children can readily follow such a process of metamorphosis in their imagination, and the thoughts invoked by the "picture" lead to the realization that the diagonals of a rhombus bisect one another at right angles. Such a metamorphosis through movement is a most valuable exercise for the developing thought life of the child. There is, of course, a corresponding metamorphosis starting from a rectangle and ending with a square (Figs. 42 and 44).

Fig. 47 shows a very interesting configuration based on what we have called "the fundamental circle drawing" (Fig. 4). The whole page is covered with this basic design, and then a network of parallel lines is drawn joining the centers of the circles vertically, horizontally, and diagonally. It will be seen that, with one exception, all the three- and four-sided figures we have dealt with so far are to be found in this "framework"; the square is the only one that is not to be found. Further, this fundamental circle design also illustrates some important properties concerning circles, one of which is shown: A tangent to a circle is at right angles to the radius at the point of contact. Other circle properties are shown in the following drawing (Fig. 48), which is a small section of the basic circle "framework" with the appropriate straight lines drawn in:

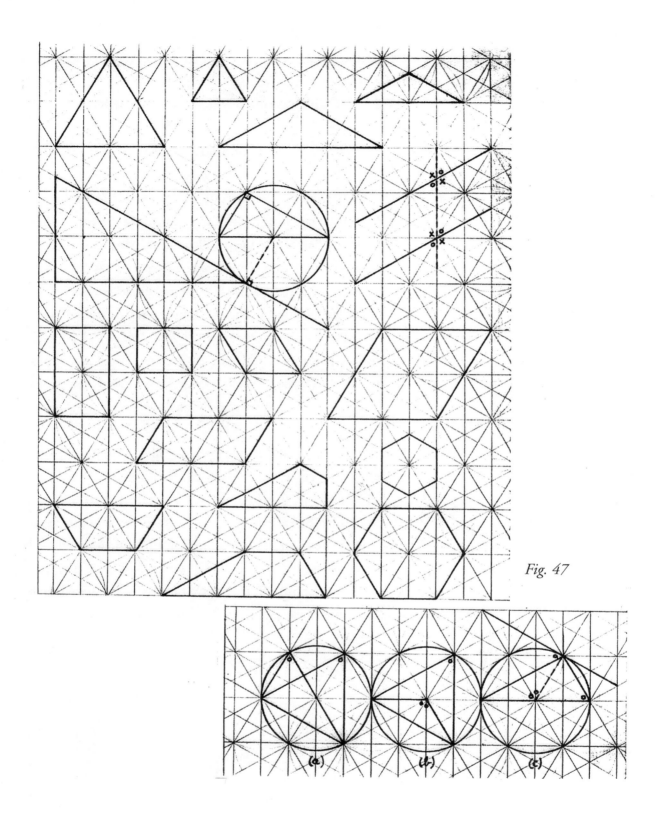

Fig. 47

Fig. 48 a, b, c

Circle (a) : Angles in the same segment of a circle are equal.

Circle (b) : The angle at the center of a circle is double the angle at the circumference standing on the same arc (also illustrated in circle (c)).

Circle (c) : The angle between a tangent to a circle and a secant is equal to the angle in the alternate segment.

As well as the discovery of individual geometrical figures and their laws and properties within this network of circles and straight lines, circle drawing also forms a basis for the most varied designs. One may say that almost the whole of elementary geometry is contained in such a fundamental expression of two-dimensional space, and it is a fine educational experience for children at the commencement of their studies in geometry to discover the variety of forms that naturally belong to this space. The drawing of the framework itself is also a fine exercise for care and accuracy and gives the children great pleasure in finding how the circles and lines weave together with such wonderful harmony and symmetry. We shall have cause to refer to this again in the later chapter on projective geometry.

As a kind of culmination to these studies of the simple geometrical figures, we may deal with one of the most important theorems in geometry: the theorem of Pythagoras. The demonstrations of the truth of this theorem that we shall now consider involve certain facts about triangles, squares, and parallelograms that we have already considered. The well-known statement is as follows: **The area of the square on the hypotenuse of a right-angled triangle is equal to the sum of the areas of the squares on the other two sides.** In his great work, *Mysterium Cosmographicum* (1596), Kepler speaks of two things in geometry as of the first importance about which everyone should know: the theorem of Pythagoras and the geometry of the pentagram. (To this second theme the next chapter will be devoted.) There are many different proofs and demonstrations of the truth of the theorem of Pythagoras. One of the earliest is Chinese, from the Book of Chou Pei Suan King, probably written about A.D. 40, but considered to come from a source before the time of Pythagoras. And five or six thousand years ago, the Egyptians and Babylonians used a special case of this law in their method of "squaring the temple." That is, they knotted together three pieces of rope of lengths proportional to 3, 4, and 5 and pegged this down on the ground at the knots, thus obtaining a right angle. The proof of this law best known in the Western world is that of Euclid (Euclid I.47), in which we have a logical argument based on a number of other proofs that have gone before. In our own school days, many of us were intellectually convinced of the truth of Pythagoras, but we probably had no real experience that the area of the two smaller squares together made up the area of the large one. And to know something with the head only is to half know it or perhaps, even, to not really know it at all. At the age we are here considering—12 or 13 years old—it is of the greatest importance that children should gain a real experience of what they are learning by seeing it and doing it. Then later, when children are 15 or 16, the intellectual proof based on logical argument can rightly come as a kind of complement; then they will "know" with their whole human nature and not merely with a part of themselves. The following drawings illustrate two different demonstrations that the children can carry out and then see for themselves that the large square is equal in area to the sum of the other two. The first is a "cutting out" exercise and is shown for an isosceles right-angled triangle and then for any scalene right-angled triangle (Figs. 49 and 50). The second is a "movement" demonstration (Fig. 51a, b, c, d, and e), and to understand Fig. 51d, the children must, of course, have experienced the truth of the proposition that parallelograms on the same base and between the same parallels are equal in area. The teacher may therefore prefer to leave this demonstration until this fact has been dealt with.

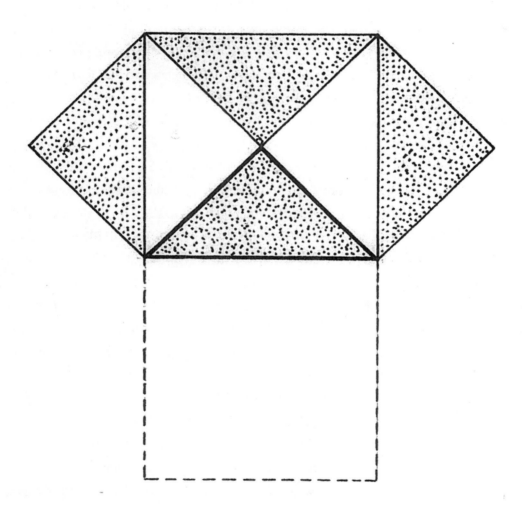

Fig. 49

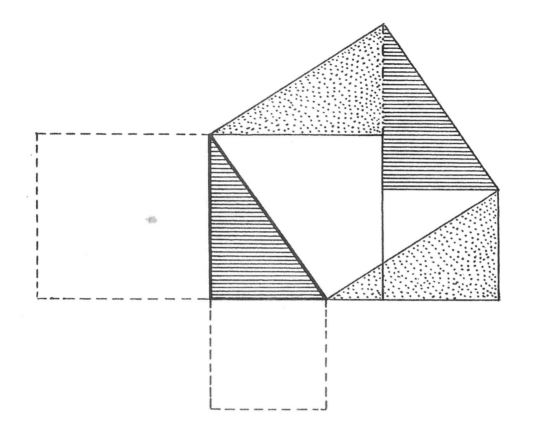

Fig. 50

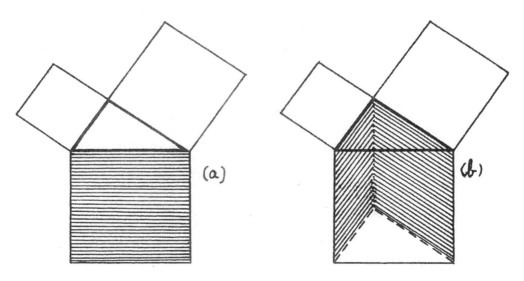

Fig. 51 a, b

59

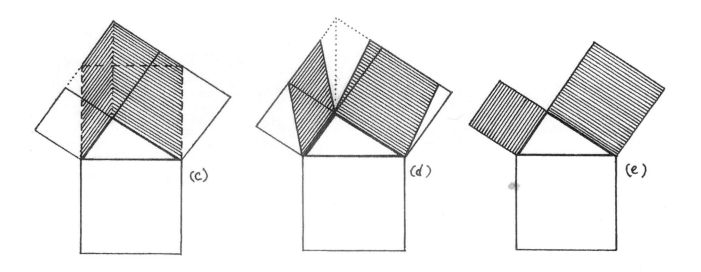

Fig. 51 c, d, e

Several times during our consideration of the triangle, areas have been referred to, but so far we have not needed to know how to calculate the area of a triangle. It has simply been a question of seeing that certain figures have the same area because they are the same shape and size. Here it is assumed that children of this age are familiar with the concept of area and know how to find the areas of rectangular figures. In this connection it is interesting to note that once upon a time, the value of land was probably reckoned by how much corn or rice could be grown upon it. This method could be also adopted to show the truth of Pythagoras by sprinkling sand evenly over the three square "fields," then collecting the sand from the two smaller fields into a little heap, and seeing that this was the same size as the heap collected from the large field. Obviously, one could only get an approximate result, but such a procedure does give children a real feeling for what the area of a surface signifies.

In the introductory chapter, reference was made to the manifold forms of nature and their relations to pure geometrical forms, and it was pointed out that it should be a vital part of education in this subject to lead children to a realization of the fact that geometrical laws are at work in the creation of natural forms. A beautiful example of geometrical form in nature based on the numbers 5 and 6 is to be seen in the fossils of sea lilies or crinoids. Figs. 52a and b are sketches of the two sides of such a calcareous fossil drawn full size; it is slightly convex on one side (Fig. 52a) and concave on the other and shows a five-petaled flower form, wonderfully symmetrical, while the whole cellular structure of the fossil is hexagonal. Sea lilies are sessile echinoderms related to starfish, brittle stars, sea cucumbers, and so on. They are thus starfish that have been turned upside down and affixed to the bed of the sea by stalks of varying length.

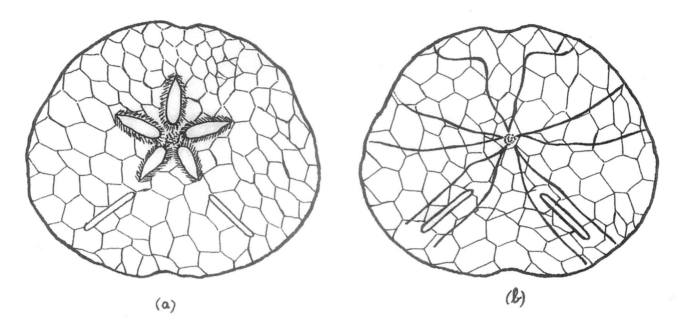

Fig. 52 a, b

In the next series of drawings showing the continuous metamorphosis of a triangle according to a quite definite process, we see how the straight-line form of the triangle passes over into curves that are clearly similar to well-defined leaf forms. Rays are drawn from points in the sides of an equilateral triangle to the center of the triangle (and beyond it as necessary). Then each point in the sides is moved along its own particular ray towards the center the same distance; this construction is clearly shown for one side of the triangle in Fig. 53a where none of the points has reached the center. In Fig. 53b the midpoints of each side have reached the center, and we get three curves, each with a cusp, the whole giving a very characteristic leaf form. In Fig. 53c some points have passed through the center and some have not, and each side then becomes a curve with a loop in it. The three angular points of the triangle have reached the center in Fig. 53d, and all other points have passed beyond the center: a trefoil form results. In Fig. 53e, all the points have gone through the center of the triangle. The construction, therefore, for this series of drawings is very simple, although the joining of the points obtained by a smooth curve will present difficulties to some children. Such curve drawing will, however, become increasingly important in later work, and it is a good exercise for children to undertake it at this stage. Artistic coloring of these drawings, that is, shaded coloring, which is related to the movement towards and beyond the center of the triangle, adds greatly to the interest and pleasure that the children have in constructing the geometrical forms. This principle of construction may be extended to other geometrical figures, and modifications may be introduced to produce curves reminiscent of other well-defined leaf forms.

The next two drawings illustrate the metamorphoses of a pentagon and of a circle with such a modification of the construction introduced. The resulting curves clearly resemble an ivy leaf and a water-lily leaf. In the case of the pentagon (Fig. 54) rays from points in the sides are drawn to the center of the figure. (At this stage the pentagon is drawn by measuring the angle 108°. In a later chapter dealing with the geometry of the pentagon, the construction in a circle is described.) The point 0 is not moved at all, and the point 10 is moved along its ray all the way to the center while the two points 1 are moved $1/10$ of this distance, the two points 2 are moved $2/10$ of this distance, and so on. These varying distances moved by the different points are given by the perpendiculars in the accompanying construction diagram in which the length of the base line is equal to the semi-perimeter of the pentagon. For the circle (Fig. 55), the construction for the water-lily leaf is just the same: The base line in the accompanying construction figure is equal to the length of the semi-circumference of the circle.

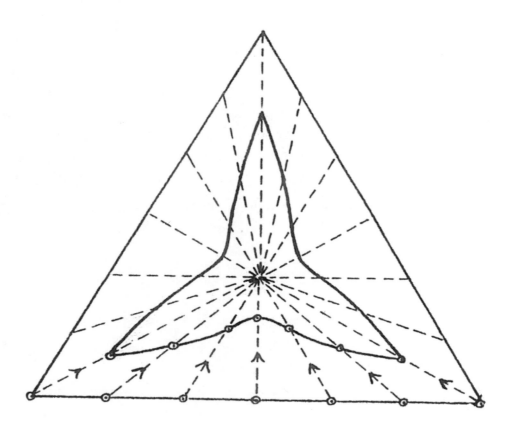

Fig. 53a

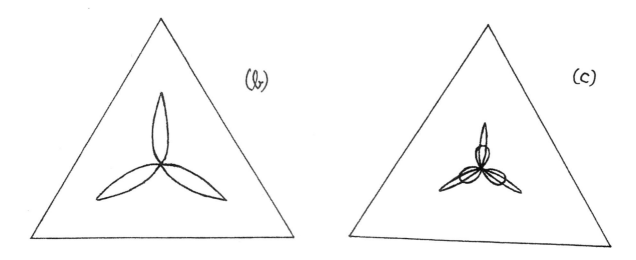
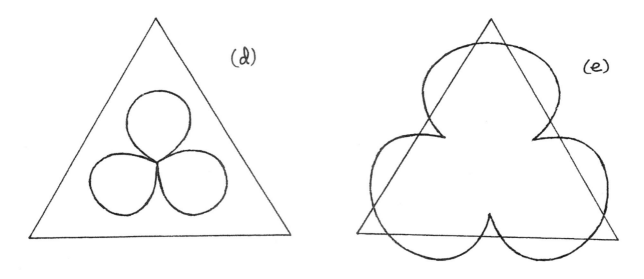

Fig. 53 b, c, d, e

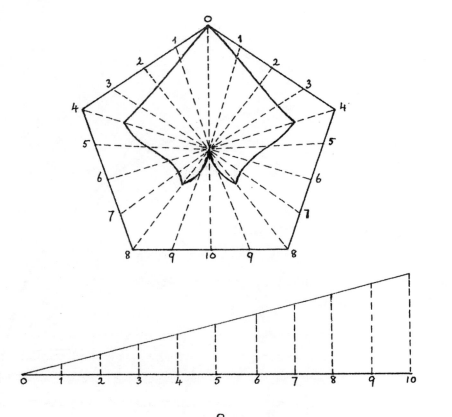

Fig. 54

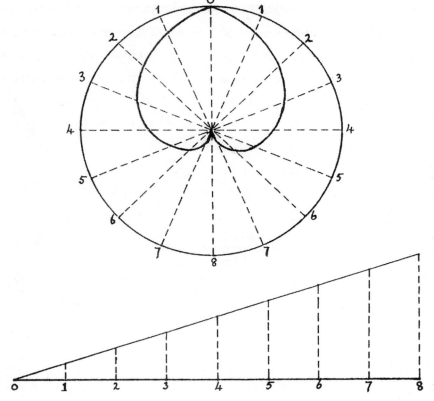

Fig. 55

64

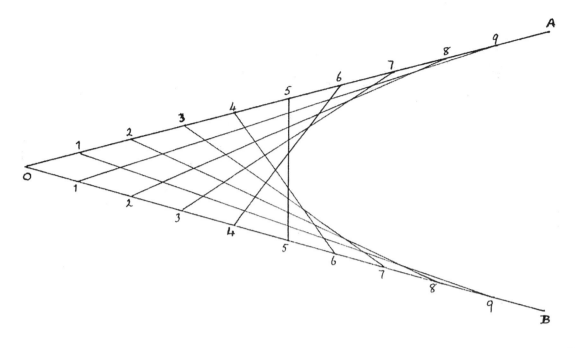

Fig. 56

In the last series of drawings illustrating the metamorphoses of a triangle, a pentagon, and a circle, we have arrived at characteristic leaf-form curves, which have been drawn freehand by joining points obtained according to definite geometrical construction. Such metamorphoses were originally developed by Dr. H. von Baravalle to whom reference has already been made in the preface. This "point-wise" method of drawing a curve is, of course, a very common one and belongs essentially to analytical or Cartesian geometry—to the drawing of graphs. There is, however, another means of constructing a curve, which children of this age should certainly practice. This is the "line-wise" method and is really a kind of molding of the curve from the outside. We have already come across this construction in Fig. 19 where the weaving of many straight lines molds a number of concentric circles—the straight lines form the envelopes of the circles. There now follow examples of such line-wise-constructed curves. Mrs. Edith L. Somervell introduced line-wise curves, early in the twentieth century, in the teaching of quite young children by using cards punched with different series of holes through which were threaded colored silks or wools (see Chapter 1.) The parabola is the easiest and simplest curve to construct, and Fig. 56 shows the line-wise construction. Two straight lines, OA and OB, are drawn at any angle, and equal distances are marked off along these lines and numbered as shown. Then points 1, 2, 3, and so on, in OA are joined to points 1, 2, 3, and so on, in OB. We see at once how the curve arises, and it is evident that the more points we take along each line, the "smoother" the curve will appear. The curve is, of course, really made up of straight lines that envelop the curve. Fig. 57 illustrates the same construction using lines OA and OB at right angles and taking many more points along the lines; here we see a much smoother curve. Mrs. Somervell recommends that the children should only draw (or stitch) this curve after they have been introduced to the "curve of pursuit." Here one imagines a rabbit feeding at O some distance from his burrow at A (Fig. 58). A dog at B sees the rabbit and gives chase. The rabbit makes a "beeline" for his burrow while the dog must always be changing his direction and so runs along the curved path as shown. This curve of pursuit is not a parabola though it somewhat resembles one; it belongs to a class of curves known as transcendental. In this illustration the dog is traveling at the same speed as the rabbit, and so the rabbit safely "goes to earth" since the dog takes a longer route!

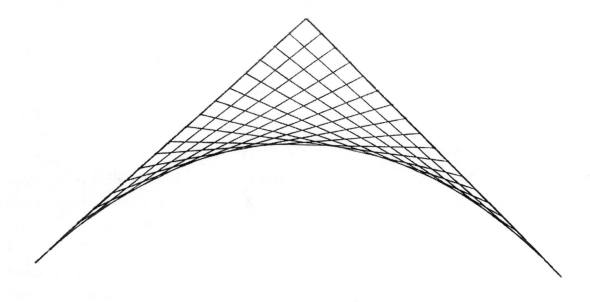

Fig. 57

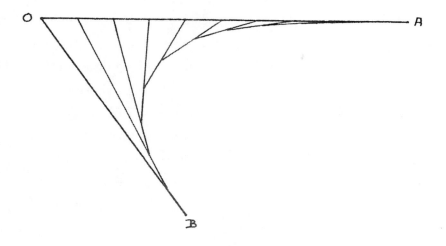

Fig. 58

Another important curve that may easily be constructed in a line-wise manner is the logarithmic spiral (Fig. 59). Here a series of 30°, 60°, and 90° triangles are placed one on the other as shown, and the short sides of these triangles envelop the curve. It may be noted that by joining the angular points (where the triangles come together) by a smooth curve, we obtain a point-wise logarithmic spiral.

This important spiral is also a curve of pursuit. The problem of three dogs placed at the vertices of an equilateral triangle, and starting simultaneously with equal velocities to chase one another, leads to the logarithmic spiral as the curve of pursuit for each dog.

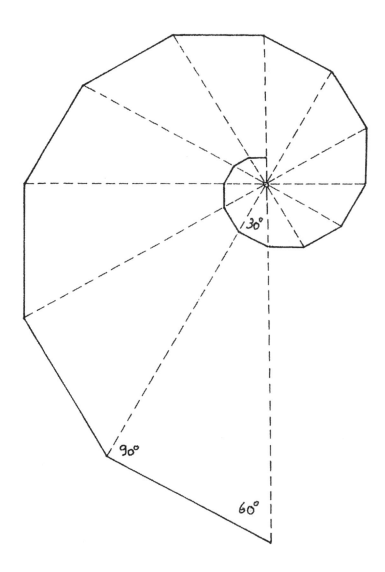

Fig. 59

In their later studies, the children will come across both these curves in quite other connections: the parabola, when they learn about the conic sections—Chapter VIII—and the logarithmic spiral in connection with the golden ratio—Chapter IV. It is an excellent opportunity when a teacher can refer to material of such importance with which the children have had some experience at an earlier age.

There are endless possibilities for drawing beautiful designs by the line-wise construction of curves. Figs. 60, 61, 62, and 63 show examples using a triangle, a square, and a hexagon; notice that all these particular designs involve combinations of parabolic curves.

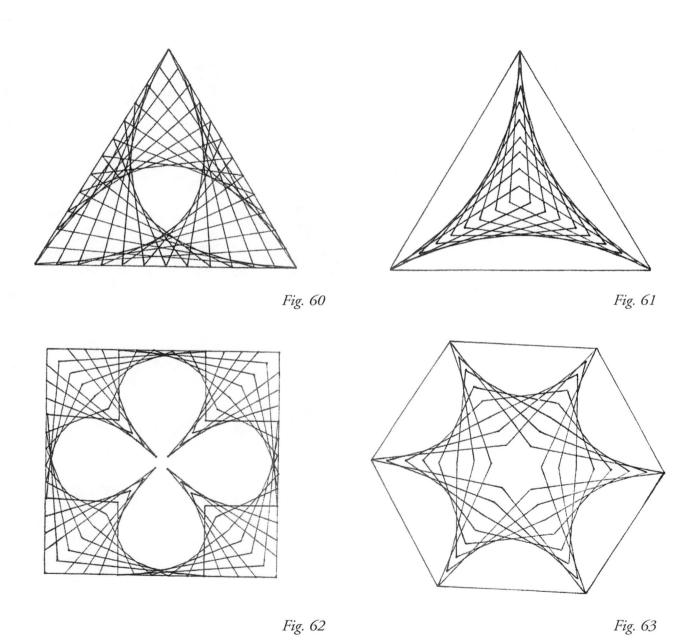

Fig. 60

Fig. 61

Fig. 62

Fig. 63

The examples that have been given in this chapter have mostly been taken from the writer's experience of the teaching of geometry to younger children, say from 12 to 14 years of age. Such a course would cover about two years' work, and during this time the children will gain a sound knowledge of the elementary geometrical figures together with their laws and properties, a considerable amount of practice in the fundamental constructions of geometry, an introduction to the drawing of curves, as well as some feeling for the relationship of geometry to form in nature. All this has been given to the children in an imaginative, pictorial way, that is, in a manner that belongs to their particular stage of development. Later on when they treat the subject more intellectually with the Euclidean proofs, they will already have a firm foundation on which to build, and this different approach will then give them a knowledge complementary to that which they acquired in these early years.

It should also be emphasized that in using such methods, the teacher has great scope for her own imagination and inventiveness, and the foregoing must in no way be considered as a complete syllabus. Different teachers will choose different examples and perhaps stress one aspect of the work more than another. The essential is that the subject shall be introduced by means of pictures appealing to the imagination and not by trying to force the intellect prematurely.

It is not the intention of this book to deal exhaustively with every part of a school syllabus in geometry. In the chapters that follow, examples will be given (1) to show what may be a new approach to ordinary themes included in most school curricula and (2) to introduce aspects of geometry that are perhaps unusual in schoolwork but which the author believes are essential to an understanding of form in nature and of the nature of space itself.

Chapter 4

The Pentagon and Pentagram, and the Golden Ratio

In the last chapter, reference was made to the statement of Kepler that the two most important things to learn about in geometry are the theorem of Pythagoras and the geometry of the pentagram. In the following study of the pentagon and pentagram, we shall see why Kepler considered these figures of such importance.

The two figures, of course, belong together: the regular five-sided figure, the pentagon, and its re-entrant counterpart, the pentagram, or five-pointed star. We have already drawn a pentagon in the series of metamorphosis drawings (Chapter 3). We will now repeat the drawing and include the pentagram (Fig. 64), constructing it by again measuring the angle of 108°. (We shall see later why the angle is 108°.) From very early times, this five-pointed star has had great significance as a symbol, and we shall realize as we study it that it is indeed a true picture of man himself with his arms outstretched, his feet firmly planted on the earth, and his head pointing to the heavens. It is no mere arbitrary picture, but a real symbol of the human being in his threefold nature. In Greek times, the pentagram was sometimes called the "triple triangle." Iamblichus, in his writings about the Pythagorean School of philosophy and mathematics, tells us that the pentagram was the sign and seal of that school. It was considered by the Greeks as symbolical of health, and probably the star points were denoted by the letters of the word $υψελα$—our word *hygiene*—the diphthong $ελ$ being replaced by "$θ$" (see Fig. 64).

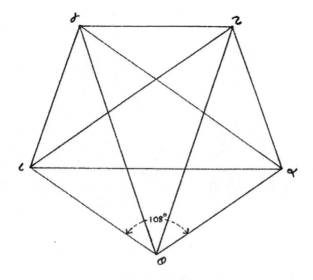

Fig. 64

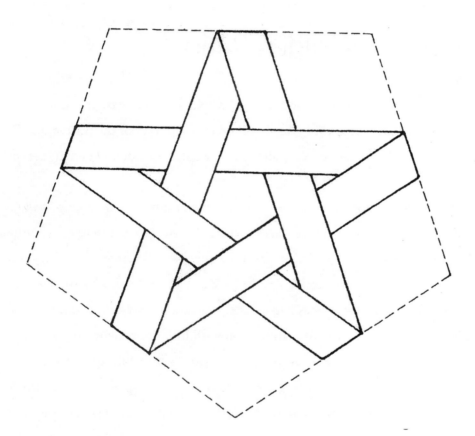

Fig. 65

There is the story of one of the pupils of the Pythagorean school of mathematics who was on a journey; he fell ill and sought shelter at a wayside inn. In spite of the care of the innkeeper he became worse, and, realizing that he was going to die, he called the host to him, thanked him for his care and devotion, and regretted that he had no money with which to pay him. He asked the innkeeper to bring him a board on which he drew a pentagram, telling him to hang it outside the inn. Soon after this he died, and the innkeeper hung the board with the pentagram as his inn sign. Years afterwards, another Pythagorean, travelling that way, noticed the sign and asked the innkeeper how he came by it. On hearing the host's story, this traveler then paid him generously for the hospitality and attention he had given to the dying man.

The pentagram has always been used as a Christian symbol, for example, the star at the top of the traditional Christmas tree, while the six-pointed star is the emblem of the Jewish faith. The pentagram is a single broken line that returns to its starting point; the hexagram consists of two separated equilateral triangles, one pointing to the heavens and the other to the earth. In the Middle Ages, the pentagram was often worn as an amulet or charm to ward off evil: The Devil would flee before the pentagram just as he would at the sign of the Cross. This mystical significance of the figure is referred to by Goethe in one of the early scenes of *Faust*: Mephistopheles enters Faust's study in the form of a poodle, which changes by magic into human form. After a short conversation with Faust, Mephistopheles asks permission to leave him:

Meph.: Pray let me leave you for the present.

Faust: I do not see why thou shouldst pray.
Though our acquaintance be but recent,
Look in upon me day by day.
Here is the window, there the entrance,
A chimney I can offer you.

Meph.: Let me confess—there is a trifling hindrance
Which bars my course the doorway through
The wizard's foot upon your threshold.

Faust: The Pentagram! That gives thee pain?
If that thy foot within the mesh hold,
Thou Son of Hell, how didst thou entrance gain?
Say, how was such a spirit cheated?

Meph.: Observe it well! That figure's not completed;
Here, if you look but closely, it remains
A little open at this outer angle.

Faust: A lucky chance, the Devil thus to entangle!
So thou'rt my captive for thy pains?
Nay, by my fay, that is a windfall!

Meph.: The poodle leapt across it all unmindful,
But now things wear another face!
The Devil cannot void the place.

(from Goethe's *Faust*, translation by Albert G. Latham)

Another striking reference to the symbolism of the pentagram is to be found in one of the Arthurian legends:

Gawain was known for a good knight, faithful in five ways and five times in each way. He was like refined gold, pure from any vileness and radiant with all virtues. Therefore he bore the pentangle as his emblem, as the truest and gentlest of all the knights. First he was faultless in his five wits; and then he never failed in the might of his hands and the skill of his five fingers. He put all his trust in the five wounds that Christ bore on the cross. And whenever he stood in the press of fight he kept steadfast in his mind, through all the tumult, that he drew all his might in battle from the five joys that the gracious Queen of Heaven had of her child. For this reason he had, on the upper half of the inside of his shield, a picture of the Virgin painted, so that when he looked at it his courage never failed. And the fifth five that Gawain had were the five virtues: generosity and love of his fellow men and cleanness, and courtesy that never failed and lastly pity, that is above all other virtues; these five were deeper in Gawain's heart and more surely part of him than of any other knight. With these five he was girded and each was joined with the others. There are five fixed points in the pentangle, and no line runs into another nor yet is sundered from the rest; and there is no place wherever a man begins, at which he can come to an end of the figure. Therefore on Gawain's bright shield the device was charged splendidly gold upon gules. This is the pure pentangle as wise men call it.

(*Sir Gawain and the Green Knight* by M. R. Ridley)

The pentagram or pentangle referred to above would be emblazoned on Sir Gawain's shield in the form of a continuous band or ribbon as shown in Fig. 65.

The mathematics and geometry of the pentagram now to be described are suitable for children of 14 or 15 years of age. We may begin, apparently far away from the subject, by considering different series of numbers, starting with the simpler series of the arithmetic and geometric progressions. The children should experience the qualitative difference of these two series: how the arithmetic series (e.g., 2, 4, 6, 8, 10, 12, 14, etc.) increases (or decreases) in a regular, even manner, while the numbers of a geometric series (e.g., 2, 4, 8, 16, 32, 64, 128, etc.) increase (or decrease) by "leaps and bounds" with greater and greater rapidity the further we go in the series. We may say that the first series proceeds in a rather dull and sluggish way, while the second one has great activity within it. Here again, reference may be made to earlier work, and the children may be reminded of two drawings they did in their first geometry lessons where the lengths of the lines and the areas of the figures (triangles in the one drawing and squares in the other) decrease in geometrical progressions (Chap. 3, Figs. 11 and 20). It will also be possible at this stage to introduce the well-known methods of finding the nth terms of these series and also the summation of n terms. Here the story of little Karl Friedrich Gauss (1777–1855) may be told: He had a very lazy schoolmaster who one day set his small pupils (Gauss was 6 years old at the time) the task of adding together all the numbers from 1 to 100, thinking that this would keep them occupied long enough for him to have a nap! He dozed off at his desk but after a few moments felt a gentle tug at his coattails. Looking down, he saw a very small boy holding a large slate on which was written just a few figures and the correct answer to the sum he had set!

1,	2,	3,	98,	99,	100
100,	99,	98,	3,	2,	1
101,	101,	101,	101,	101,	101

10,100 divided by 2 = 5,050

Such stories of mathematical development and "discovery" are important for children to hear, for they bring real human interest into the subject as well as illustrating mathematical genius.

The children will greatly enjoy making up more or less difficult number series, and then others in the class can try to find out how they have been built up and what are the various mathematical processes involved in their formation. Eventually the teacher will write the following series on the chalkboard:

0, 1, 1, 2, 3, 5, 8, 13, 21, 34, 55, 89, 144, 233, etc.

This progression is built up in such a way that each term is the sum of the two preceding terms. It is the well-known Fibonacci series, named after Leonardo Fibonacci of Pisa, who had a great reputation as a mathematician during the early years of the thirteenth century and who introduced the use of Arabic numerals into Christian Europe.

We now take successive terms of the Fibonacci series, expressing them as ratios and finding their decimal values:

$0/1$	=	0	$21/34$	=	0.61764....
$1/1$	=	1	$34/55$	=	0.61818....
$1/2$	=	0.5	$55/89$	=	0.61797....
$2/3$	=	0.66666....	$89/144$	=	0.61805....
$3/5$	=	0.6	$144/233$	=	0.61802....
$5/8$	=	0.625	$233/377$	=	0.61803....
$8/13$	=	0.61538....	$377/610$	=	0.61803....
$13/21$	=	0.61904....	etc.		etc.

We see that the different ratios are alternately greater and less than a certain number (0.61803....) and that they approximate more and more closely to this number the further we go in the series. This process may be compared with the vibration of a pendulum, swinging first to one side and then to the other, each swing being less than the one before it and all the time getting nearer and nearer to its mean position but never quite "dying down."

Another series is now built up in the same way as the previous one, and again the successive terms are expressed as ratios with their decimal values: For example:

1, 3, 4, 7, 11, 18, 29, 47, 76, 123, 199, 322, 521, 843, etc.

$1/3$	=	0.33333....	$76/123$	=	0.61788....
$3/4$	=	0.75	$123/199$	=	0.61809....
$4/7$	=	0.57142....	$199/322$	=	0.61801....
$7/11$	=	0.63636....	$322/521$	=	0.61804....

$11/18$ =	0.61111....	$521/843$ =	0.61803....
$18/29$ =	0.62868....	$843/1364$ =	0.61803....
$29/47$ =	0.61702....	$1364/2207$ =	0.61803....
$47/76$ =	0.61842....	etc.	etc.

Again the same thing happens: Out of quite a different set of ratios a certain number gradually appears, and this number is the same as for the previous series, that is, 0.61803 .... So we may start with any two numbers, and from them build up a Fibonacci series, and the ratios thus obtained will always more and more closely approximate this same strange number.

An exact mathematical expression for this number to which the Fibonacci series ratios approximate is:

$$\frac{\sqrt{5}-1}{2} = 0.6180339885....\text{(to 10 decimal places)}$$

This number has an endless number of decimal places without any recurring. That is, it is an irrational number because $\sqrt{5}$ is irrational—and it has very special and remarkable properties. For example, if we calculate its reciprocal, that is, divide it into unity, we obtain a whole number (unity) followed by exactly the same decimal digits.

Thus:

$$\frac{\sqrt{5}-1}{2} = 0.6180339885...$$

$$\frac{2}{\sqrt{5}-1} = \frac{1}{0.6180339885...} = 1.6180339885...$$

The proof of this remarkable property is as follows:

$$\frac{2}{\sqrt{5}-1} = \frac{2}{\sqrt{5}-1} \times \frac{\sqrt{5}+1}{\sqrt{5}+1} = \frac{2(\sqrt{5}+1)}{4} = \frac{\sqrt{5}+1}{2} = 1 + \frac{\sqrt{5}-1}{2}$$

From the foregoing, it is clear that this is a very special, even unique number, and we will therefore now call it the golden number and denote it by the letter G. Another way of expressing G is by the following continued fraction:

$$G = 1 + \cfrac{1}{1 + \cfrac{1}{1 + \cfrac{1}{1 + \cfrac{1}{1 + \text{etc.}}}}}$$

If we stop this continued fraction at different points and simplify it, we always get a ratio whose denominator and numerator are consecutive terms of the Fibonacci series. Thus in the above case, the fraction simplifies to 13. Again, another expression for G is the following continued square root:

$$G = \sqrt{1 + \sqrt{1 + \sqrt{1 + \sqrt{1 + \sqrt{1 + }}}}} + \text{etc.}$$

This expression is, of course, more difficult to simplify, involving the use of logarithms. (It should be pointed out that the letter G is used either for the number 0.61803 . . . or its reciprocal 1.61803 . . . as may be convenient. In the above two expressions, the value is 1.61803 . . . .)

As well as forming the golden number, the numbers of the Fibonacci series have many remarkable properties among themselves. Dr. J. Ginsburg of Yeshiva University, U.S.A., has pointed out a number of such relationships among which are the following:

The reciprocal of 89, the twelfth term of the Fibonacci series, itself gives in its decimal equivalent the Fibonacci series:

$$1/89 = 0.011235955056179\ldots$$

```
            0.0112358
                   13
                    21
                     34
                      55
                       89
                       144
                        233
                         377
                          610
                           987
                           1597
            _____
            0.011235955056179.......
```

Again, if any three successive terms of the Fibonacci series are taken, the sum of the cubes of the two greatest ones less the cube of the smallest one always gives a Fibonacci number.

$$(F_{n+2})^3 + (F_{n+1})^3 - (F_n)^3 = \text{Fibonacci number}$$

For example, $\quad 5^3 + 3^3 - 2^3 = 125 + 27 - 8 \quad = 144$

or, $\quad 8^3 + 5^3 - 3^3 = 512 + 125 - 27 = 610$

The value of G (this time, 0.61803 . . . .) is also given by the positive root of the quadratic equation $G^2 + G = 1$:

$$G^2 + G - 1 = 1$$

$$G = \frac{-1 \pm \sqrt{1+4}}{2}$$

$$= \frac{\sqrt{5}-1}{2} \quad \text{(positive value)}$$

$$= 0.61803\ldots$$

Another remarkable property of the golden number is shown by raising the number to the second, third, fourth (and so on) powers:

$$G = 1.618 \text{ (correct to 3 decimal places)}$$
$$G^2 = 2.618 \text{ (correct to 3 decimal places)}$$
$$G^3 = 4.236 \text{ (correct to 3 decimal places)}$$
$$G^4 = 6.854 \text{ (correct to 3 decimal places)}$$

etc.   etc.

From the above we see that

$$G^2 = G + 1 = 1 + G$$
$$G^3 = G^2 + G = 1 + 2G$$
$$G^4 = G^3 + G^2 = 1 + 3G$$

Thus, $G^5 = G^4 + G^3 = 3 + 5G$

and $G^6 = G^5 + G^4 = 5 + 8G$

etc.   etc.   etc.

In the third column of the above table, the numbers of the Fibonacci series appear again. We also see that we have here arrived at a Fibonacci series, which is at the same time a geometrical progression with the common ratio G:

1. $G, G^2, G^3, G^4, G^5 \ldots\ldots\ldots G^n, G^{n+1}, G^{n+2}$, etc.

Thus, $G^n + G^{n+1} = G^{n+2}$

or, $1 + G = G^2$

i.e., $G^2 - G - 1 = 0$

$$G = \frac{1 \pm \sqrt{5}}{2}$$

$= 1.61803\ldots.$ (considering positive values only)

It must be remembered that the common ratio G of the above series is itself a ratio arrived at ever more and more closely by taking successive terms of the Fibonacci series. This series, which can be obtained by adding two successive terms to get the next one or by multiplying each term by a common ratio, is quite unique and again emphasizes the remarkable properties of the golden number.

In his well-known book, *The Curves of Life* (Constable & Co. Ltd., 1914), Sir Theodore Cook deals in considerable detail with the golden number and its properties using a somewhat different notation from the above. Whereas here we have used the letter G as the symbol for the golden number or its reciprocal as is convenient, Cook uses the symbol $\Phi = 1/G$. The equation thus becomes:

$$\frac{1}{\Phi^2} + \frac{1}{\Phi} = 1$$

i.e $$\Phi^2 - \Phi - 1 = 0$$

which is the above equation where we have used the symbol G.

So far we have considered certain number relationships and have discovered a number that has very remarkable and unique properties, a very special number—truly a golden number. It should be mentioned here how such a study of the quality of number can give children the greatest interest in mathematics. They realize that there are wonderful secrets hidden within numbers, and they see how often the simplest arithmetical processes reveal these secrets. Mathematics is more than a science of calculation for utilitarian purposes, which in ordinary schoolwork it so much tends to become. Children can and should experience the quality of numbers, and they can then gain some understanding of the Greek conception of number. Aristotle, referring to the Pythagorean School of mathematics, said:

> They thought they found in numbers, more than in fire, earth, or water, many resemblances to things which are and become; thus such and such an attribute of numbers is justice, another is soul and mind, another is opportunity and so on; and again they saw in numbers the attributes and ratios of the musical scales. Since, then, all other things seemed in their whole nature to be assimilated to numbers, while numbers seemed to be the first things in the whole of nature, they supposed the elements of numbers to be the elements of all things, and the whole heaven to be a musical scale and a number.

A good example of this qualitative treatment of number is given by the Greek conception of a "perfect" number or again of a "friendly" number. A perfect number is one whose factors (including unity which the Greeks always considered as a factor) add together to make the number. For example, 6 and 28 are perfect numbers:

$$6 = 1 + 2 + 3 \qquad \text{and} \qquad 28 = 1 + 2 + 4 + 7 + 14$$

Pythagoras was once asked by one of his pupils concerning the nature of friendship; he replied that friends are to one another as the two numbers 220 and 284. To understand this strange answer, one only has to see that the sum of the factors of 220 (including unity) is 284 and the sum of the factors of 284 is 220.

Sum of factors of 220 = 1 + 2 + 4 + 5 +10 +11 + 20 + 22 + 44 + 55 +110 = 284
Sum of factors of 284 = 1 + 2 + 4 + 71 + 142 = 220

There are, of course, other perfect numbers and pairs of friendly numbers.

    To return then to our subject after this digression, we shall now see that what we have dealt with here is really the mathematics of the pentagram, and we will therefore consider the geometry of this figure. The pentagram may be drawn by two methods, first by construction in a circle and then by measurement of sides and angles. The method of construction for drawing a pentagram in a circle is carried out in the following steps (see Fig. 66): Draw a circle, center O, and place in it a diameter XOY and a radius OA at right angles. Bisect OY, giving point P. With center P and radius PA, draw an arc cutting the diameter in Q. Then distance AQ (stepped off with compasses) will go exactly five times round the circle, giving points A, B, C, D, E. If now these points are joined in their order, a pentagon is obtained, and if alternate points are joined, we get a pentagram.

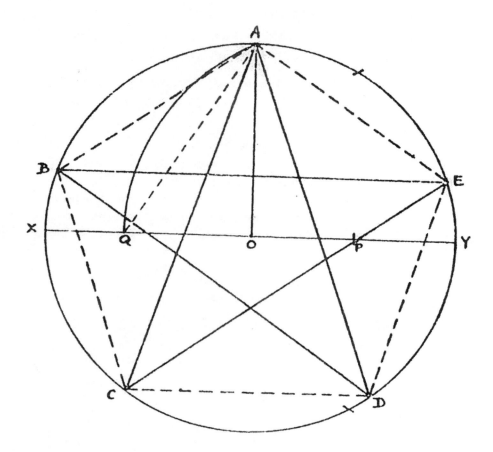

Fig. 66

The second method of drawing the figure by measurement involves knowing the angle of a regular pentagon, namely 108° (see Fig. 64). This angle may be calculated by applying the Euclid proposition, which says that all the angles of any rectilinear figure, together with four right angles, are equal to twice as many right angles as the figure has sides. (The proof of this proposition may well be given to the children.) Thus if θ is the required angle of a regular pentagon, then

$$5\theta + 4 \text{ right angles} = 10 \text{ right angles}$$
$$5\theta = 6 \text{ right angles}$$
$$= 540°$$
Therefore, $\theta = 108°$

In all the foregoing work on the mathematics of the pentagon-pentagram figure, the children will, of course, carry out the calculations involved. This will give them valuable practice in elementary arithmetic, for example, in the finding of the decimal values of the Fibonacci series ratios. Now when they have learned to construct the figure accurately, they will be able to draw many designs based on it. One such design is the pentagram band or ribbon (Fig. 65). Three further examples are shown in Figs. 67, 68, and 69, in which a decagon is first constructed in a circle. The coloring of such designs can bring out different characteristics. For example, in Fig. 68 there are very many pentagrams to be found. Fig. 67 is, of course, a continuous decagram band.

Fig. 67

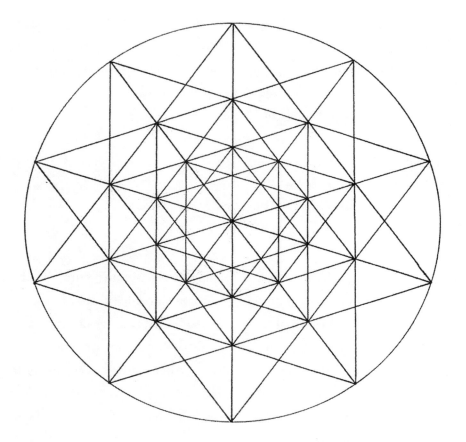

Fig. 68

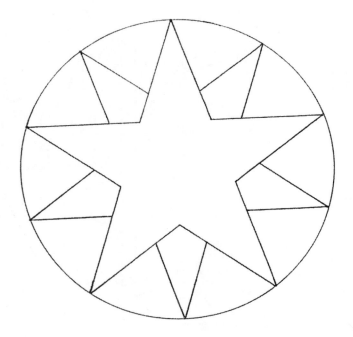

Fig. 69

84

Now it is evident that all the sides of the pentagram are of equal length and that the figure is symmetrical with respect to a line drawn from an angular point through the center. There being five such lines, the pentagram has a fivefold symmetry. Each of its sides is parallel to a side of the enclosing pentagon, and thus there are five such pairs of parallel lines giving rise to five equal trapeziums. ABCE is one such trapezium (Fig. 70). There are also a number of rhombus figures, for example ABPE (Fig. 70). Various geometrical forms can thus be found within the pentagon-pentagram figure: other trapeziums (BEQT, BRQT, SRDC), other rhombuses (BSQT), quadrilaterals (ASVR), and many triangles, all of which are isosceles. A very important and interesting property is that the figure reproduces itself, for we can go on drawing pentagrams one within the other, each one being smaller than the previous one and also being alternately erect and inverted.

We may now consider the angles of the pentagon-pentagram. Take the angles grouped round point P: the smallest angle is TPX = angle XPY = angle YPQ. Let us call this angle α. Then angle TPQ = 3α. Now, angle XPQ = 2α = angle QPD. So we have:

$$3\alpha + 2\alpha = \text{angle TPQ} + \text{angle QPD} = 2 \text{ right angles}$$

That is, $5\alpha = 2$ right angles $= 180°$

Therefore, $\alpha = 36°$

So the angles grouped round point P are as follows:

angle TPX	=	36°
angle TPY	=	72°
angle TPQ	=	108°
angle TPD	=	180°
reflex angle TPC	=	288°

and we see that these angles are in the proportion of the numbers 1, 2, 3, 5, 8, which we recognize as the first few terms of the Fibonacci series.

As we have seen, the triangles to be found in the figure are all isosceles but of two kinds, acute angled and obtuse angled, a "tall" form and a "flat" form (Fig. 71). The tall form has a special angle property in that its base angles (72°) are double the vertical angle (36°); also the vertical angle (36°) of the tall form becomes the base angle of the flat form whose vertical angle is obtuse (108°).

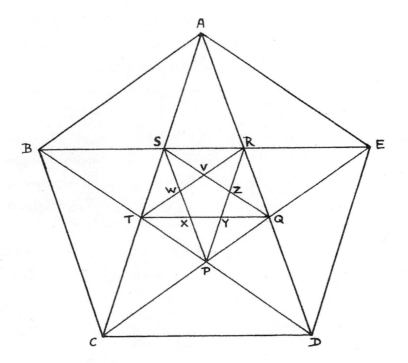

Fig. 70

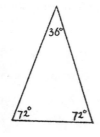 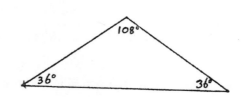

Fig. 71

Lastly we have to consider the lengths of the lines in the pentagon-pentagram figure, and the following series gives the lengths, increasing from the smallest to the greatest to be found in Fig. 70. (Of course, a longer series of lengths could have been obtained if further pentagrams had been drawn, but the figure as it stands—three pentagons and two pentagrams—gives enough for our purpose.)

1st	Length of side of	smallest pentagon	-	say, VZ
2nd	Length of star-point	smallest pentagram	-	say, SV
3rd	Length of side of	middle-sized pentagon	-	say, SR
4th	Length of side of	middle-sized pentagram	-	say, SQ
5th	Length of side of	largest pentagon	-	say, AB
6th	Length of side of	large pentagram	-	say, AC

Consider the 3rd length, SR:
    We see that SR = SZ = SV + VZ
    i.e., the 3rd length = 2nd length + 1st length.

Consider the 4th length, SQ:
    We see that SQ = VQ + SV = SR + SV
    i.e., the 4th length = 3rd length + 2nd length

Consider the 5th length, AB:
    We see that AB = BR = BS + SR = SQ + SR
    i.e., the 5th length = 4th length + 3rd length.

Consider the 6th length, AC:
    i.e., the 6th length = 5th length + 4th length.

Thus we see that this series is again a Fibonacci series, for each length is the sum of the two preceding ones. But it is a very special Fibonacci series, for if we express successive pairs of terms as ratios as was done before, we find that the value of these ratios no longer approximates closer and closer to the golden number the further we go in the series, but that every single ratio gives the golden number exactly! This may be shown by measuring the lengths as accurately as possible, and the value of each ratio will then be, say 1.62 (correct to two decimal places). The fact that consecutive lengths belong to triangles that are similar also proves that the ratios are all equal.

There are many different ways in which a line may be divided in geometry to give certain proportions, and among them there is the well-known proportion of the golden ratio:

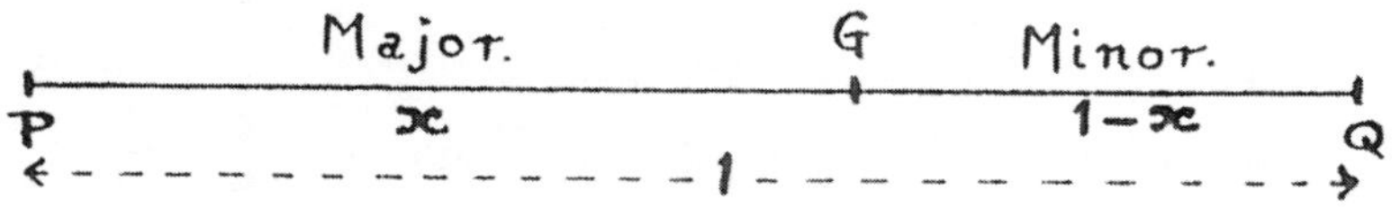

Fig. 72

The line PQ (Fig. 72) is divided in golden ratio at the point G if the following condition is fulfilled: The ratio of the smaller part (minor) to the larger part (major) equals the ratio of the larger part (major) to the whole line.

i.e., $$\frac{\text{Minor part}}{\text{Major Part}} = \frac{\text{Major part}}{\text{Whole line}}$$

$$\frac{GQ}{PG} = \frac{PG}{PQ}$$

Such a condition is fulfilled many times in the pentagon-pentagram figure. For example, take two similar triangles, next to one another in size, such as ACD and ATQ. Because they are similar, their corresponding sides are proportional:

i.e., $$\frac{TQ}{CD} = \frac{AT}{AC}$$

But, $CD = AT = AB$

Thus we have $$\frac{AS}{AB} = \frac{AB}{AC}$$

and these two equal rations are formed from consecutive lengths AS, AB, AC, which we have shown belong to a Fibonacci series.

It now only remains to show that the golden number is the numerical expression of the golden ratio: If we consider the whole line to be divided in golden ratio as of unit length and the large part of length x, then the small part is of length (1 − x). Then we have, according to the definition of this proportion, the condition that

$$\frac{1-x}{x} = \frac{x}{1}$$

i.e., $\quad 1 - x = x^2$

or $\quad x^2 + x - 1 = 0$

which is the same equation we considered earlier, and the positive root is

$$X = \frac{\sqrt{5}-1}{2} = 0.61803\ldots$$

Thus, $\quad \dfrac{\text{Minor part}}{\text{Major Part}} = \dfrac{\text{Major part}}{\text{Whole line}} = \text{the golden number}$

To sum up, we see that every pair of consecutive lengths in the pentagon-pentagram figure expressed as a ratio gives us the golden number. This ratio occurs again and again in the figure. Indeed we may say that the pentagram is built upon the golden ratio; the whole figure is "saturated" with it. It can also be shown that certain areas in the figure stand to one another in golden ratio, for example, the areas of pairs of consecutive triangles, the triangles being arranged in order of magnitude (See Fig. 70).

Thus, $\quad \dfrac{\text{area tri SWV}}{\text{area tri SVR}} = \dfrac{\text{tri SVR}}{\text{tri RSW}} = \dfrac{\text{tri RSW}}{\text{tri RSQ}} = \dfrac{\text{tri RSQ}}{\text{tri ASR}} =$

$\dfrac{\text{tri ASR}}{\text{tri SBA}} = \dfrac{\text{tri SBA}}{\text{tri ATQ}} = \dfrac{\text{tri ATQ}}{\text{tri ABE}} = \dfrac{\text{tri ABE}}{\text{tri ACD}} = \text{the golden number}$

For example, consider the last pair of ratios.

Then area tri ATQ = 1/2 base × height $\quad$ = 1/2 TQ × height

Then area tri ABE $\quad$ = 1/2 AB × height

Then area tri ACD $\quad$ = 1/2 AC × height

Now the heights of these three triangles onto the bases TQ, AB, CD are equal. Let us call these equal heights h.

Then we have $\dfrac{1/2\,TQ \times h}{1/2\,AB \times h} = \dfrac{1/2\,AB \times h}{1/2\,AC \times h}$

i.e., $\dfrac{TQ}{AB} = \dfrac{AB}{AC}$

but TQ, AB, and CD are consecutive lengths in the pentagon-pentagram figure.

Therefore, $\dfrac{TQ}{AB} = \dfrac{AB}{AC} =$ the golden number

Therefore, $\dfrac{\text{area tri ATQ}}{\text{area tri ABE}} = \dfrac{\text{tri ABE}}{\text{tri ACD}} =$ the golden number

It is very interesting to notice the different names that have been given to this geometrical proportion. Pacioli (c. 1450–1510) called it *Proportio divina*; Kepler (1571–1630) referred to it as *Sectio divina* (divine proportion or ratio), and Leonardo da Vinci (1452–1519) gave it the name of *Sectio aurea* (golden ratio). Today if this proportion is looked up in any school geometry, it will be found under the name of "medial section" or extreme and mean ratio. Such a name, which has no real meaning, gives us no feeling for the importance and the quality of this proportion, which earlier mathematicians and artists certainly considered it to have.

The following is the geometrical construction for dividing a given line in golden ratio:

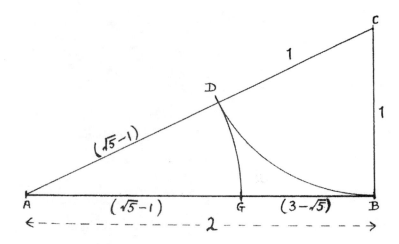

Fig. 73

If AB is the given line (Fig. 73), then at one end, say B, draw a perpendicular BC making BC = $1/2$ AB. Join AC. Then with center C and radius CB, draw an arc cutting AC in D. With center A and radius AD, draw an arc cutting AB in G. Then the point G divides line AB in golden ratio.

Proof of the above construction:

Suppose AB is 2 units of length. Then by construction, BC = 1 unit and CD = 1 unit.

By Pythagoras, $\qquad AC = \sqrt{5}$

Therefore: $\qquad AD = \sqrt{5} - 1$

Therefore by construction: $\qquad AG = \sqrt{5} - 1$

Therefore: $\qquad GB = 2 - (\sqrt{5} - 1) = 3 - \sqrt{5}$

Now for line AB to be divided in golden ratio at G the condition is that

$$\frac{\text{Minor part GB}}{\text{Major part AG}} = \frac{\text{Major part AG}}{\text{Whole line AB}}$$

Minor part GB = $\dfrac{3-\sqrt{5}}{\sqrt{5}-1} = \dfrac{3-\sqrt{5}}{\sqrt{5}-1} \times \dfrac{\sqrt{5}+1}{\sqrt{5}+1} = \dfrac{2\sqrt{5}-2}{4} = \dfrac{\sqrt{5}-1}{2} = \dfrac{\text{Major part AG}}{2}$

Therefore, AB is divided in golden ratio at G.

Using this construction, we may now draw a golden rectangle, that is, a rectangle whose sides are in golden ratio (Fig. 74). It is characteristic of this figure that if we successively cut off squares on the smaller side, we get a series of similar—that is, always "golden"—rectangles shrinking in "golden series" along one of the diagonals. Any slight inaccuracy of proportion is quickly revealed by the increasing "oblongness" of the diminishing squares.

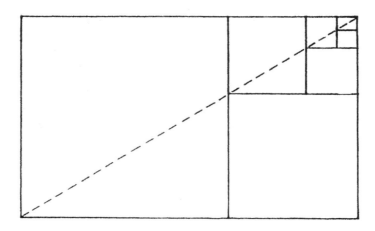

Fig. 74

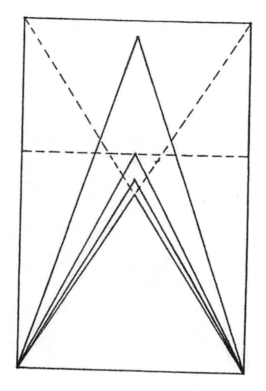

Fig. 75

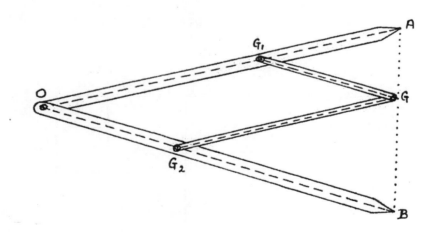

Fig. 76

The drawing of Fig. 75 shows a comparison of angles. The four triangles are drawn inside a golden rectangle. The lowest isosceles triangle is given by the diagonals of the golden rectangle. It is a slightly flattened equilateral triangle—base angles just under 60° and vertical angle just over 60°. The second triangle is equilateral—angles 60°. The third is given by the semi-diagonals of a square—base angles 63°26', vertical angle 53°8'. The fourth is the pentagram triangle—base angles 72°, vertical angle 36°. It may be noted that the approximation of the lowest triangle to an equilateral triangle gives a convenient method for a rough-and-ready construction of a golden rectangle and therefore of golden ratio.

Fig. 76 is a diagram of "golden compasses," which may be used for measuring golden ratios. They may be made of wood or metal. The two long equal legs OA and OB, pivoted at O, are divided in golden ratio at G_1 and G_2. The two shorter legs, G_1G and G_2G, joined and pivoted at G, are also pivoted at G_1 and G_2, respectively, and are equal in length to the minor and major lengths of the golden ratio division of the two longer legs. To whatever distance the compasses are stretched, AGB is always a straight line (the pivot at G should be as near the end of each leg as possible), and the point G then divides the distance AB in golden ratio.

It is interesting to note that a similar instrument, a pair of double-ended dividers (hinged like a pair of scissors) was discovered during the excavations at Pompeii and was evidently used in these early times for setting out the golden ratio to any scale.

From the very earliest historical times we find many references to this important proportion and to its mathematical expression, the number G. We have already pointed out that the pentagram was the symbol of the Pythagorean School of mathematics, and it may well be that Pythagoras learned about this figure and its proportions from the Egyptian priests with whom he studied. That this proportion was well known to the Egyptians and that they considered it of the first importance is clear from the fact that it is expressed quite fundamentally in the measurements of the Great Pyramid at Giza. The inclination of each of the triangular faces to the square base is given by J. H. Cole of the Survey Department of the Egyptian Government (1925) as 51°52'. He also gives the side of the square base as 756 ft (actually this is the average of the measurement of the four sides, which differ by only a few inches) and the vertical height as 481 ft. From these two measurements, we may easily calculate (by the theorem of Pythagoras) that the height of each of the face triangles is 612 ft. Now the cosine of the angle 51°52',

i.e., $\dfrac{\text{Half the side of the base}}{\text{Height of face triangle}} = \dfrac{378}{612} = 0.618 = G$ to three decimal places.

(Cos 51°52' = 0.6180)

Thus with an astonishing degree of accuracy in such a massive structure, the golden ratio is embodied. It is also interesting to notice that the cotangent of the angle 51°52'

$$= \frac{\text{Half the side of the base}}{\text{Vertical height of pyramid}} = \frac{378}{481} = 0.786 = \frac{\pi}{4}$$

(cot 51°52' = 0.7860)

Many Greek temples were built in golden ratio proportion, not only in their overall measurements but in fine details as well. The overall breadth and height are generally major and minor. The main door is the major to the height of the pillars. In the total height of the building, the height of the pillars is the major and the upper part (entablature and pediment) the minor. The entablature and pediment are minor and major to one another. The width of triglyph and metope are often in golden ratio. In many cases, the distance between the two middle pillars is the major and that between the other pillars the minor. Also the distance between pillars and their thickness is sometimes in golden ratio. The door is a golden rectangle. These are a few examples among many that could be given. Never perhaps have the beauty and grace of proportion of a Greek temple been surpassed, and we see how these qualities are just those expressed by the golden ratio. Of course this proportion is embodied in many other styles of architecture both before and after Greek times and in different epochs of civilization. It is indeed universal and for all time.[2]

In his book *Der Goldene Schnitt*,[3] Dr. Goeringer speaks of how in handicrafts the true craftsman is striving to make something aesthetically beautiful as well as practical and useful, and he points out that the proportions of such creations often embody the golden ratio. He chooses two extreme examples of craftsmanship, the American ax and the violin, and in the following drawings (Figs. 77 and 78), he indicates the proportions of these two forms, illustrating how intimately they are bound up with the golden ratio, although they are used for such widely differing purposes. The violin is such a perfect example of the craftsman's art and it embodies the golden ratio so fundamentally that we will consider the geometry of its form in some detail. Recently the author was given a drawing showing the geometrical constructions used today in the famous Mittenwald School of violin making in Germany. Fig. 79 shows most of the essentials of this drawing reproduced to half-scale. (The body of a violin is about 14 inches long.) It shows the proportions of the widths of different parts of the body in relation to the length. There are two fundamental points that determine these proportions. These are the geometrical center C, the midpoint of the line of symmetry AB, and the position of the bridge D. This point D is found as follows: BB_1 is perpendicular to AB and half its length (i.e., $BB_1 = BC$). Join B_1A, and along this line step off $B_1C_1 = BC$. Then D is the foot of the perpendicular from C_1 on to AB. The position of point R, through which passes the line of greatest width, is determined by dividing CB in golden ratio by the construction as shown; the position of point P is then found by stepping off CP = CR. The position of the feet of the bridge is given by the circle with center D and radius DC. The second circle is twice the radius of the first one and determines the position of point Q, through which passes the line of least width at the "waist" of the violin. Point X is fixed by dividing AD in golden ratio, and then point Y, by dividing BX in golden ratio. The actual widths of the body at these different fundamental points are found as follows: half the narrowest width at Q = QY; the greatest width at R is twice the narrowest width at Q; the width at P = AQ; the narrowest width at Q is also the golden ratio major of the width at P; half the width at X and the length AX form two sides of a golden rectangle; that is, they are in golden ratio; similarly for half the width at Y and the length BY. We see from these detailed considerations how the essential form of the violin is built up according to the golden ratio proportion.

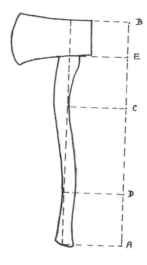

Fig. 77

$$\frac{BC}{AC} = \frac{BE}{EC} = \frac{EC}{CD} = \frac{AD}{DC}$$

$$= \frac{CF}{FA} = \frac{FD}{CF} = \frac{FD}{DA} = G$$

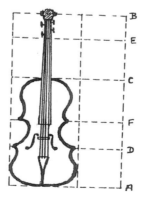

Fig. 78

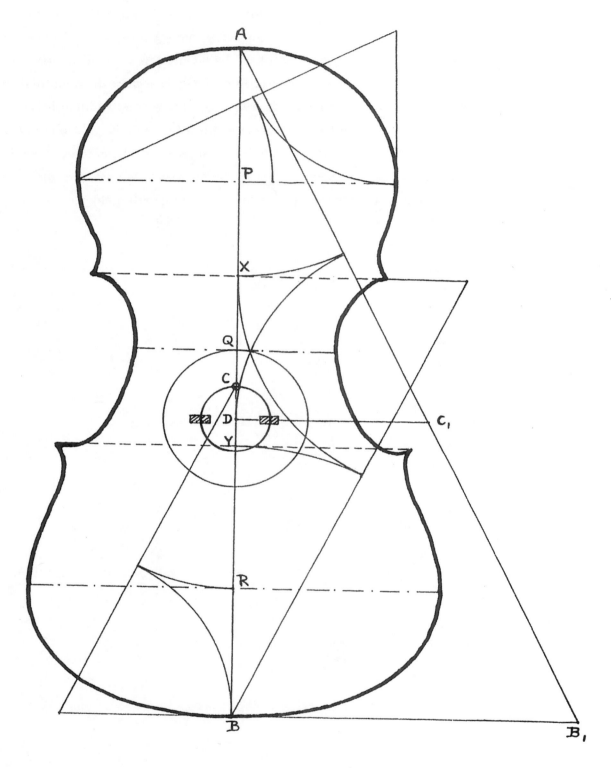

Fig. 79

The author has in his possession a small, dark blue/in which the golden ratio is to be found in all its proportions. Fig. 80 shows a full-scale plan and elevation of this vase. The diameter of the base is one-fourth of the height. The position of the greatest diameter divides the height in golden ratio (point O). The average diameter of the mouth is the minor of the greatest width of the vase, The diameter of the narrowest part of the neck is slightly less than the major of the average diameter of the mouth. On the side of the vase, beautifully executed in gold on the blue glaze, is a wild duck flying up from some reeds. The eye of the duck is exactly at the point that divides the overall height of the vase in golden ratio.[5]

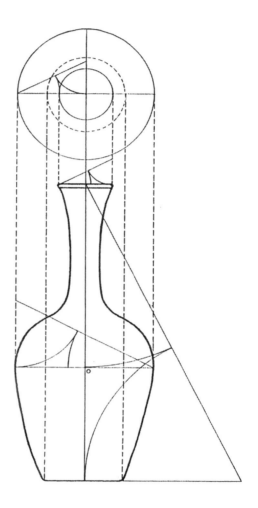

Fig. 80

Another example of the craftsman's art is the making of furniture, and here again the golden ratio is often to be found. We reproduce by kind permission of the Ministry of Works, London (Crown copyright reserved), two photographs (Plates 1 and 2) of the famous Coronation Chair in Westminster Abbey, as well as scale drawings of the back and side elevations and a plan of the seat of the chair (Figs. 81, 82, and 83). The label describing the chair itself states that it was "made by Walter, the King's Painter, at the command of Edward I in 1300–1301 to contain the Stone of Scone brought by the King from Scotland. In this chair every Sovereign has been crowned since King Edward II." The following golden ratio proportions are to be found in the chair—some of them with considerable accuracy, others very near. In each case, the ratio is expressed as major : minor (= 1.618). (The proportions to three significant figures are calculated from the elevation and plan drawings.)

<u>Back elevation</u>   (Fig. 81)

PB : AB = 1.63	TV : RT = 1.53	(thus CD = TV,
RV : CD = 1.62	UB : TU = 1.59	UB = SU,
CD : RT = 1.55	SU : QS = 1.56	and TU = QS)

Note that CD is the distance between the midpoints of the vertical pillars; also Q marks the position of the center of the circle in the triangular form at the top of the chair. The points that other letters indicate are clearly shown.

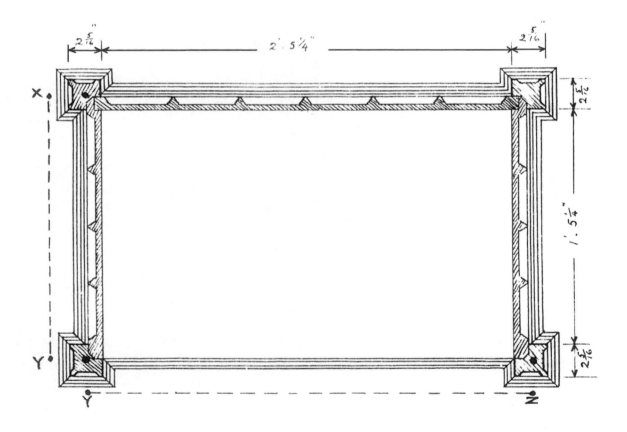

Fig. 81

Fig. 82

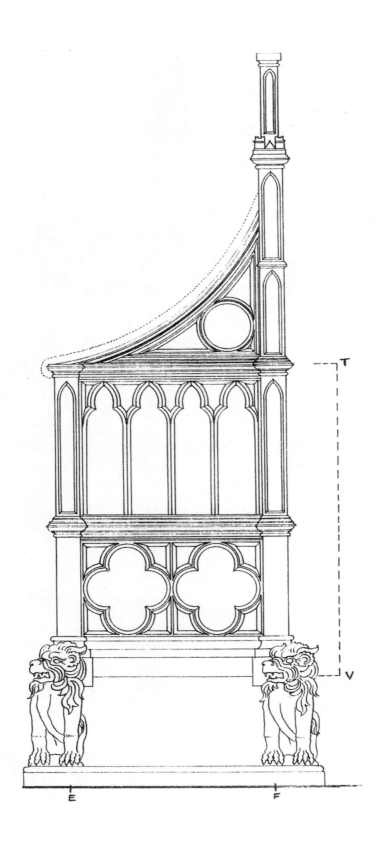

Fig. 83

Plate 1

Side elevation   (Fig. 82)

Plate 2

$$TV : EF = 1.61$$

Again EF is the distance between the midpoints of the front and back pillars.

<u>Plan of seat</u>   (Fig. 81)

$$YZ : XY = 1.62$$

The distances are taken from the midpoints of the pillars.

In a recent book on sound reproduction entitled *High Fidelity* by G. A. Briggs, the following passage appears: "Mr. J. Moir, in a lecture on Room Acoustics to the British Sound Recording Association in February 1956, said that the ideal dimensions for a listening room were: Length, 21 ft, Width, 13½ ft. and Height, 8½ ft." [6] It is interesting to notice that

$$21 : 13\tfrac{1}{2} = 1.56 \text{ and } 13\tfrac{1}{2} : 8\tfrac{1}{2} = 1.59.$$

Thus the proportions of this room, length : width and width : height are those of the golden ratio, correct to one decimal place (1.6). Here we see that this proportion belongs to what is aesthetically pleasing to the ear as well as to the eye! Indeed, Dr. Goeringer in his book, *Der Goldene Schnitt*, speaks of the golden ratio in connection with the wavelengths of tones of beautiful chords, so the ideal proportions of a listening room are perhaps not surprising.

When we look at a painting by a great artist, we are immediately aware of two things—the color and the form. We can experience at once the balance and harmony of the picture inherent in these two expressions of the artist's skill and inspiration, and we shall often find that the greatest works of art of all ages reveal the principle of the golden ratio in their form. One of the most striking examples is *The Last Supper* by Leonardo da Vinci—perhaps the greatest of all paintings (Plate 3). In the form of this picture we find the golden ratio repeatedly expressed. The "centers of gravity" of each of the four groups of Apostles are in golden ratio with the central figure of Christ; the three heads in each group, especially the faces, are arranged in this same proportion, and the general level of the heads divides the whole height of the painting in golden ratio. Other golden ratio proportions are to be found as follows: the front edge of the table, the top of the windows and the edge of the ceiling; the front edge of the table, the edge of the tablecloth, and the lower edge of the picture; the position of each small window in relation to the width of the whole painting; the width of each of the small windows and the width of the middle one; the perspective widths of the dark tapestries on the side walls.

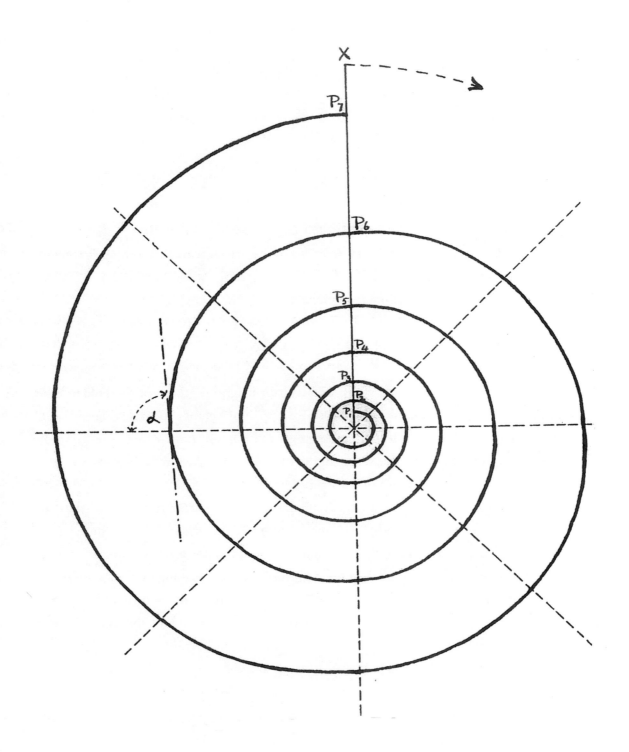

Fig. 84

Plate 3

Another striking example is Giotto's famous painting of St. Francis preaching to the birds (Plate 4). Here the position of St. Francis' head divides both the height and the width of the painting in golden ratio.

Plate 4

In his book *The Curves of Life*, to which we have already referred, Sir Theodore Cook shows the fundamental significance of the proportion of the golden ratio [7] in nature, especially in relation to spiral formations, for example, in shells, leaf formations and climbing plants, animal horns, and so on. Then he discusses through several chapters the relation of this same proportion to works of art in various spheres, for example, architecture, sculpture, and painting. In all this he goes into great detail and gives many and varied examples. Then towards the end of the book he says:

> . . . the fundamental element in that joy which the artist's creation gives us may well be the manifestation of those profound laws of nature which, in some cases, he may have deeply studied, and, in many more, he may have so instinctively appreciated that they are the unconscious motives of his style and sense of taste. If "Φ" in some way describes the principle of growth, which is one revelation of the spirit of nature, would not the artist most in touch with nature tend to employ that proportion in his work even though he were not conscious of its existence?
>
> The beauty of a shell or a flower makes an irresistible appeal to us which needs no argument. . . . The processes of growth explained by the Φ spiral, and the successive proportions they reveal, have therefore an intimate connection with the source of our pleasure in the beauty of a natural object.
>
> A great painting also makes an irresistible appeal to us which needs no argument; and I may fairly compare the masterpieces of art with the shells or flowers that have survived, because bad pictures, though they do not "die," are certainly forgotten, and need not be brought into my argument at all. . . . I have already shown that the same interesting "variations" from any such simple formula as Φ are to be found in the best art just as they are to be found in the surviving organism. It will clearly be of some significance, therefore, if I can also show that there is as great a measure of agreement with Φ in the one case as in the other. If so, it will not imply that the artist had any preconceived idea of using the Φ proportions in his composition, any more than the Nautilus had any conscious plan of developing a certain spiral in its shell. But it will suggest the possibility that there exists a very real link between those processes of artistic creation which are vaguely called "instinctive" and those principles of natural growth which are admittedly fundamental. I venture, in fact, to offer Φ as an underlying reason for what we call Beauty both in a natural object and in a masterpiece of art.

The Φ spiral to which Sir Theodore Cook refers is a logarithmic spiral whose radii vectors are in golden ratio proportion (Fig. 84). He also calls it a Pheidias spiral (hence the symbol Φ—phi—for the golden ratio) because the sculpture of Pheidias expresses the Φ proportion in great detail. A straight line OX rotates in a plane about a fixed point O, and at the same time a point P in OX moves along OX. Then the point P describes a curve called a spiral. The form of the spiral depends on the law connecting the displacement of the point P along OX with the angular displacement of OX. The length of OP at any instant is called the radius vector (t), and the angle this makes with the initial position OX is called the vectorial angle (Φ) at that instant. If the successive values of Φ are in arithmetical progression (i.e., increase in equal instants by an equal angular measurement) and the corresponding values of Φ are in geometrical progression (i.e., increase regularly by a constant ratio), then the curve traced by the point P is called a logarithmic spiral. It is also known as the equiangular spiral because of the property that the angle Φ, which the tangent at any point makes with the radius vector, at that point is a constant. Now in the logarithmic spiral drawn here, the constant ratio by which the radius vector t increases in each complete spiral turn is the golden ratio (1.61803 . . .), that is,

$$OP_2 : OP_1 = OP_3 : OP_2 = OP_4 : OP_3 = OP_5 : OP_4 = OP_6 : OP_5 = OP_7 : OP_6 = 1.61803 \ldots$$

As well as in the forms of many shells, spirals are frequently to be found elsewhere in the forms nature creates. A noteworthy example is that of the arrangement of leaves on the stem of a plant. The points on the stem at which the leaves bud forth are called nodes, and in many plants and trees, these nodes lie on a spiral curve winding round the stem. We may perhaps find, as in a cherry branch, that five leaves complete a cycle and the sixth leaf lies vertically over the first from which we started counting. Furthermore, in ascending (or descending) from the first leaf to the sixth, we have traced out two complete spiral turns. Such an arrangement is expressed by the ratio 2 : 5, that is, five leaves in a cycle consisting of two spiral turns. The botanist calls this the phyllotaxis ratio. (Phyllotaxis means leaf arrangement.) Other plants have different phyllotaxis ratios, for example, 2 : 3, 3 : 5, 5 : 8, 8 : 13, 13 : 21, and so on. Now it will be noted that the numbers that form these ratios are just the numbers of the Fibonacci series. Under normal conditions of growth, many plants give phyllotaxis ratios whose numbers belong almost invariably to the Fibonacci series. "Out of 140 plants counted by Weisse, six only were anomalous, the error thus being only four percent." Another example of phyllotaxis is in the fir cone; in some species there are five rows of "scales" spiraling up the cone in one direction and three rows winding less steeply in the other direction. Or, there may be eight rows and five rows. An American botanist named Beal examined 505 cones of the Norwegian spruce and found that 92 percent had spirals of five and eight rows; that is, only eight percent did not give numbers belonging to the Fibonacci series.

In flowers we find the same phenomenon. In a sunflower, the rows of florets are generally 34 and 55, while in a very large head they are 89 and 144. Sir Theodore Cook sums this up as follows: "The fact that plants express their leaf arrangement in terms of Fibonacci numbers, so frequently that it passes for the normal case, is the proof that they are aiming at the utilization of the Fibonacci angle which will give maximum superposition and maximum exposure to their assimilating members."[8]

The "Fibonacci angle" here referred to is the angle 137° 30' 28" and is the inverse angle of

$$\frac{\sqrt{5}-1}{2}$$ of 360°, that is, (360° − 360° × 0.61803 . . .).

This subject of phyllotaxis in relation to the Fibonacci series and the golden number has been exhaustively dealt with by Dr. A. H. Church in his *Interpretation of Phyllotaxis Phenomena,* 1920. However, it should be said that other investigators do not wholly agree with his theories and ideas (e.g., D'Arcy Wentworth Thompson in his book *On Growth and Form*).[9]

From what we have said here, it will be evident that the golden ratio is a fundamental proportion inherent in the forms of nature and in the creations of the great artists and craftsmen of all ages. Now we may ask the question: Why does the artist so often use just this proportion, and does she use it consciously or unconsciously? We know, for example, that Leonardo da Vinci was fully aware of the golden ratio, although it is inconceivable that before he started painting his masterpiece, he carried out geometrical constructions on the wall of the Refectory in Milan, dividing his canvas in golden ratio proportions. And yet we find this proportion in the picture in great detail and also with considerable accuracy. On the other hand, it may well be that Walter, the King's painter, had never heard of the golden ratio, and yet again this proportion is to be found in the famous chair he made. Is there not perhaps some deeper reason for the fact that this proportion is to be found so frequently in artistic creation?

The answer to this question has already been hinted at in the reference we have made to the sculpture of Pheidias. Born in Athens about 500 B.C., he is universally regarded as one of the greatest of Greek sculptors. Many examples of Greek statues of the age of Pheidias have been examined and shown to embody the golden ratio in great detail though, of course, with divergences and variations. Now the portrayal of the ideal human form, whether male or female, by the great Greek sculptors has never been surpassed. Is not then the golden ratio fundamental to an understanding of the proportions of the ideal human form? Sir Theodore Cook[10] considers a well-proportioned man 68 inches in height, which is F^4 if we take 10 inches as the unit of measurement $[\theta^4 = (1.618\ldots)^4 = 6.854\ldots]$. Then from feet to navel is 42 inches, or θ^3; from navel to crown of head is 26 inches, or θ^2; and from breasts to navel is 10 inches, which is the unit of measurement, or θ^{10}.

We will now show such proportions somewhat differently with reference to the female human form and will use the terminology of major and minor of a length divided in golden ratio:

Major	Minor
Foot to waist (5th or last lumbar vertebra)	Waist to crown of head
Waist to chin (or 7th [last] vertebra)	Chin to crown of head (length of head)
Crown to point of kneeling	Point of kneeling to foot
Crown of head to point where middle finger comes, with arms at side	Point where middle finger comes, with arms at side, to foot
Kneeling point to waist	Waist to chin
Chin to "sitting bone"	Sitting bone to kneeling point
Sitting bone to line under breasts	Line under breasts to chin

These are some of the chief vertical golden ratio proportions. With regard to horizontal proportions, the distance from fingertip to fingertip with arms outstretched is divided in golden ratio at either of the shoulder bones or acromions. The breadth of the shoulders (acromion to acromion) is the major to the breadth of the waist as the minor. The horizontal span of the arms equals the height of the body.

What is true of the whole body is also true of the extremities. For example, the ideal hand expresses the same fundamental proportion:

Major	Minor
Tip of middle finger to center of palm	Center of palm to wrist
Wrist to knuckle	Knuckle to first finger joint
Knuckle to first finger joint	First finger joint to second finger joint
First finger joint to second finger joint	Second finger joint to fingertip

These proportions apply to each finger. The thumb where the two phalanges are nearly equal is an exception.

If the arm is held horizontal, then from acromion to fingertip is divided in golden ratio at the elbow joint. When the arm is hanging by the side, then the division is not at the elbow but just above, at the thinnest part of the upper arm where the muscles begin.

If we consider the whole height of the head from chin to crown, then the golden ratio division is at that point in the forehead known as the glabella, behind which lie the pituitary body and the pineal gland. It may be noted here that the vanishing point of all the lines of perspective in Leonardo's picture of *The Last Supper* (Plate 3) is just this point in the forehead of the central figure of Christ. Taking the height of the face only, then the golden ratio division is at the root of the nose (major below, minor above) or at the point of the nose (major above, minor below). In the latter proportion, the major is also equal to the breadth of the face.

What we have said here will be sufficient indication that the ideal human form is "built up" according to the golden ratio proportion even down to the greatest detail. In reality the ideal form is never fully attained in any individual—it is a prototype towards which every single human form more or less approximates. "So God created man in His own image, in the image of God created He him; male and female created He them."

Here then, surely, is the explanation of why the golden ratio is to be found so universally in the forms of artistic creation! The artist or craftsman projects outwardly into his work what belongs to him inwardly in his own form and nature. And we, who admire his work, find it aesthetically right and pleasing for just the same reason, although we may have no conscious knowledge of the golden ratio proportion.

We began this chapter by considering the geometry of the pentagram. The pentagram is indeed a true symbol of the human form, for as we have seen, it is built up on the golden ratio.

Chapter 5

The Four Rules of Arithmetic

The series of drawings that follows illustrates the four rules of arithmetic treated geometrically. This is a period of work that can well be carried out by children of 14 or 15 years old and introduces them to important and interesting curves, some of which they will be studying in greater detail when they are older. It also involves an understanding of the concept of point locus, and, in carrying out the drawings, much good practice is gained in the free-hand drawing of curves as the path of a moving point. Furthermore, it will be seen how the curves become metamorphosed and what their limits are. It should again be pointed out that careful and artistic coloring adds greatly to the appearance of the drawings, and the children then have more enjoyment in their work. This in no way detracts from the necessity of accurate construction and fine curve drawing.

Addition

A certain convenient length (in the drawings reproduced here, 4 inches) is chosen as the "constant sum" and remains constant for this set of drawings. Two points, called foci, are taken quite close together (say 2 inches apart) (Fig. 85a) and remain fixed for the particular drawing. Now a point is found such that the sum of its distances from the two fixed foci is equal to the constant sum (4 inches). This is done by "stepping off" these focal distances from the two foci with a compass. This process is repeated until a whole series of points is obtained, and these are then joined by a smooth curve. Thus the sum of the focal distances of any point on this curve is a constant (4 inches):

$$a + b = c + d = \text{constant (4 inches)}.$$

This closed curve, which is a geometrical picture of addition, is an ellipse. Now what happens when the foci are moved further apart, the constant sum (4 inches) remaining the same? As may be seen from the drawings (Figs. 85b and 85c), the ellipse becomes thinner, and it becomes clear that there are definite limits. When the foci are coincident, we shall have a circle (radius 2 inches). Then as they are separated, we get ellipses that become ever thinner and thinner until when the foci are a definite distance apart (in the case we are considering, this distance is 4 inches), the ellipse has degenerated into a segment of a straight line of length 4 inches. So in this process of movement, we have passed from a circle, through a family of ellipses, to a straight line. It should be noticed that the length of each ellipse—the major axis—is always the same, and equal to the constant sum (4 inches).

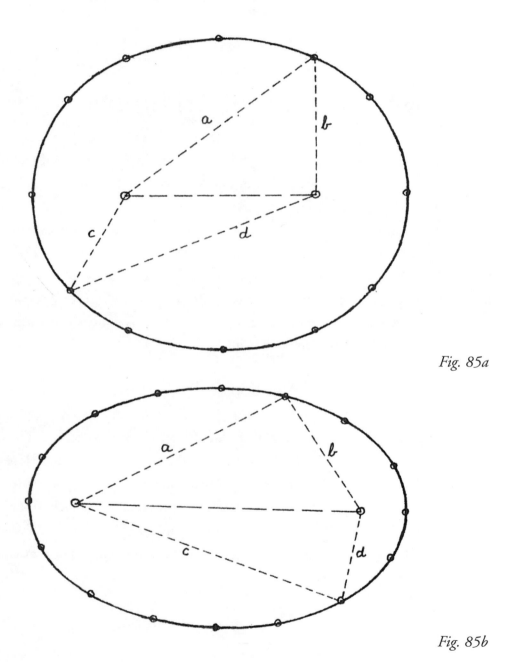

Fig. 85a

Fig. 85b

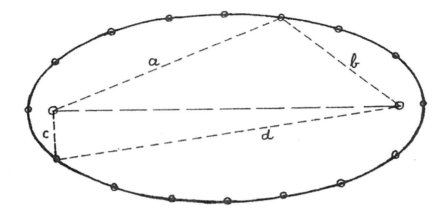

Fig. 85c

Subtraction

Again we choose any convenient length as the "constant difference" (e.g., 2 inches) and take two focal points. Now it is a question of finding all points in the plane of the paper, the difference of whose distances from the two foci is always the same, that is, equal to the constant difference (2 inches). The construction is carried out as before, but the foci must be a certain minimum distance apart before any figure can be constructed. This minimum distance is equal to the constant difference (here 2 inches). The curve obtained when this distance is greater than 2 inches is the hyperbola (Figs. 86a, b, c) whose two branches open out more and more as the foci are moved further apart. Thus a family of hyperbolas is obtained and the difference of the focal distances of any point on the curves is a constant (2 inches):

$$a - b = c - d = \text{constant (2 inches)}.$$

As may be seen from the metamorphosis of the form of the hyperbola in the three drawings, the two branches or "wings" of the curve will, in one instance, close up into two straight line segments going outwards from the foci. This limiting case will occur when the foci are the distance apart equal to the constant difference (2 inches). The other limit is when the foci are infinitely far away in either direction; then the wings of the hyperbola will have opened out to two parallel straight lines. Hence during this movement of the foci, we have passed through a metamorphosis of a hyperbola from the two outer segments of a straight line (later in this book we shall understand that this is really one continuous segment joining the foci through infinity), through the ever-widening wings of the hyperbolic curve, to two parallel straight lines at right angles to the direction of the former line. It should also be noticed that the two poles of the hyperbola (i.e., the two points where the curve cuts the line joining the foci) do not move and that their distance apart is equal to the constant difference (2 inches).

In the case of the ellipse, the functions of the two foci are the same, since for the process of addition the commutative law is valid, that is, a + b = b + a. Thus the geometrical picture of addition is a closed curve. But since for the process of subtraction this law does not hold (i.e., a − b does not equal b − a), the functions of the two foci of the hyperbola are different. In the construction of the hyperbola, one has to reverse the process of subtraction for one branch of the curve. Thus the physical reality lies in one branch or wing, and the other one is, as it were, the image or mirror picture. This is also always the case when a hyperbolic curve is obtained graphically as the expression of a physical law of nature. For example, the graphical expression of Boyle's Law,

$$PV = \text{constant},$$

where P is pressure of gas and V its volume, the temperature remaining constant, is one branch of a rectangular hyperbola. The other branch that mathematically must be there would be expressing the relation between negative pressure and negative volume. To what do such conceptions apply? Clearly not to a physical gas, which cannot have a negative volume nor exert a negative pressure!

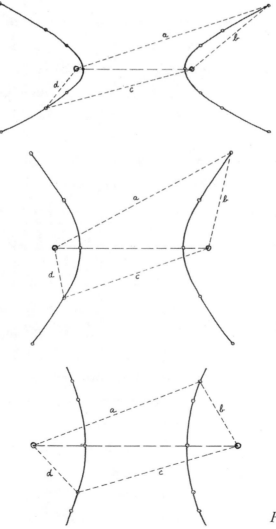

Figs. 86 a, b, c

Multiplication

In the geometrical pictures of the process of multiplication, we see a wonderful metamorphosis taking place. The length chosen this time represents the "constant product," say 4 inches, in the drawings shown in Fig. 87, and again two focal points are taken quite close together. By a similar constructional method as before, points are found, the product of whose distances from the two foci is always the same, that is, equal to the constant product (4 inches). These points are joined by a smooth curve. Then the product of the focal distances of any point on this curve is a constant (4 inches):

$$a \times b = c \times d = \text{constant (4 inches)}.$$

When the foci are quite close together (say 2 inches apart in Fig. 87a), an oval curve is obtained. This is not an ellipse. The dotted-line curve is an ellipse having the same major and minor axes as the oval. On moving the foci apart, but of course keeping the constant product the same, this oval curve becomes flattened until in one position a "flat oval" is obtained (Fig. 87b). Continuing the movement further, the flat sides of the oval become concave (Fig. 87c) until, when the foci are a certain distance apart, the figure has changed into a lemniscate (Fig. 87d). This metamorphosis occurs when the distance between the foci is equal to twice the square root of the constant product (constant product = 4, $\sqrt{4}$ = 2, 2 inches + 2 inches = 4 inches). If the foci are now further separated by only the smallest amount, two distinct egg-shaped ovals are obtained (Fig. 87e), which become smaller, more rounded, and further apart as the distance between the foci is increased (Fig. 87f). This series of curves is named after the French astronomer Jean Dominique Cassini (1625–1712), and they are known as the curves of Cassini. He proposed to substitute these oval curves for Kepler's ellipses as the paths of the planets.

By this geometrical process, we thus arrive at two distinct oval curves that have arisen from a single oval, and we have traced the various stages by which this transformation comes about. Although we see these two separate ovals arising from the single oval, they are mathematically still one curve.

A mere glance at these six drawings will remind one at once of photographs or diagrams in textbooks of biology showing the process of cell division—or rather of cell multiplication. These geometrical figures are a picture of the natural phenomenon of the multiplication of one cellular organism into two such organisms. The comparison is very striking and again shows us how geometry, treated in an imaginative and living way, may be brought into relation with organic processes of nature, with the living forms of nature that are constantly undergoing metamorphosis as they grow and develop. This relationship may perhaps be expressed by saying that underlying all form and change of form in nature there are geometrical laws at work. This is the essential theme of two well-known books, *The Curves of Life* by Sir Theodore A. Cook, already referred to in the last chapter, and *Growth and Form* by D'Arcy W. Thompson.

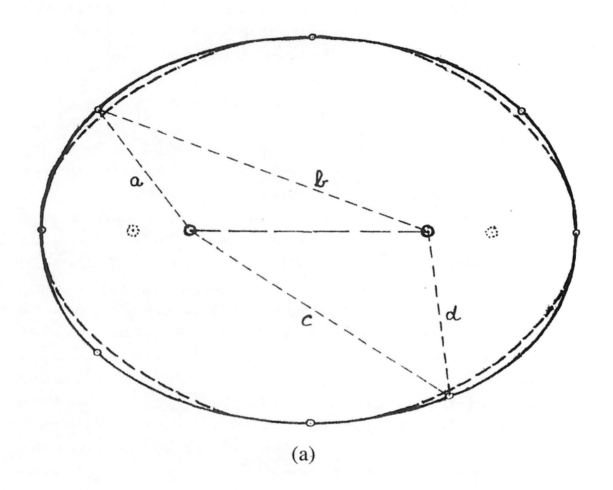

(a)

Fig. 87a

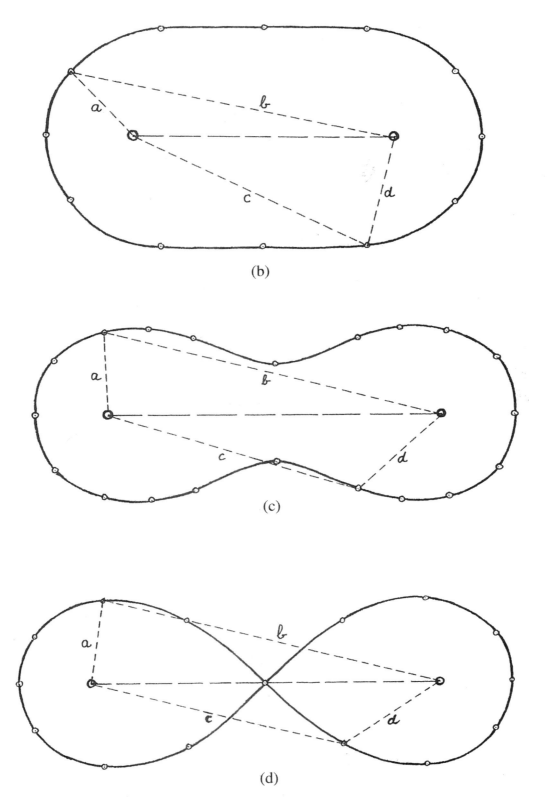

Figs. 87 b, c, d

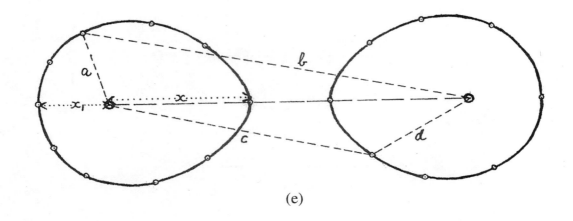

(e)

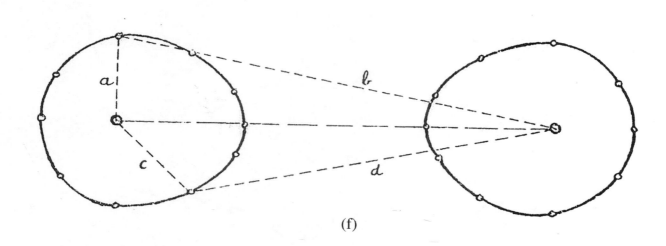

(f)

Figs. 87 e, f

In these six drawings, the product (4 inches) has been kept constant throughout while the foci have been moved further and further apart. Now we may combine them all into one picture if we keep the foci fixed and vary the product. Then each curve is a picture of the process of multiplication, the product becoming less and less as we pass from the large oval, through the flat-sided oval to the lemniscate, and on to the separated egg-shaped ovals (Fig. 88). From this figure we may obtain many interesting geometrical relationships, especially with regard to the flat-sided oval and the lemniscate. These are shown by the constructional circle and dotted lines. The flat sides of the oval are tangents to the circle passing through the two foci. A square is drawn in this circle with its diagonals; the vertices of the two equilateral triangles drawn on either side of the vertical diagonal give the two endpoints of the flat oval. Thus four important points of the flat oval are determined. Now the circle cuts two sides of each equilateral triangle, and these four points are the "highest" and "lowest" points of the lemniscate curve; the tangents to the circle drawn parallel to the sides of the inscribed square meet exactly on the endpoints of the lemniscate. So for the lemniscate, six fundamental points are determined including the "crossing point." It should be further noted that the curve crosses itself at right angles at this point.

The curves of Fig. 88 are not drawn full scale, as are those of Figs. 86 and 87, but are reproduced from a drawing in which the foci were 6 inches apart and the constant products for each curve going from outside to inside were 36, 25, 18 (flat oval), 15, 12, 10, 9 (lemniscate), 8, 6, and 4 inches. It should also be pointed out that to obtain the necessary points for drawing the curves, fractional and decimal factors of the constant product are necessary, and the endpoints are determined by the solving of a quadratic equation. To illustrate this, we will again consider Fig. 87e in which the foci are 4.1 inches apart. Here,

$$a \times b = 0.9 \times 4.5 = 4.05 \text{ and } c \times d = 3.1 \times 1.3 = 4.03.$$

Both these values of the constant product are very close to 4. To find the positions of the endpoints of the curve, that is, the distances x and x_1, the following calculation is necessary:

$$x(4.1 - x) = 4$$

$$\text{i.e., } x^2 - 4.1x + 4 = 0$$

Solving the equation, $x = 1.6$ (approximately)

$$x_1(4.1 + x_1) = 4$$

$$\text{i.e., } x_1^2 + 4.1x_1 - 4 = 0$$

Solving this equation, $x = 0.82$ (approximately).

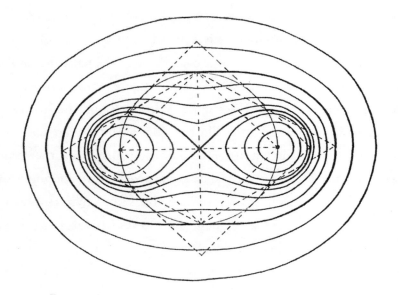

Fig. 88

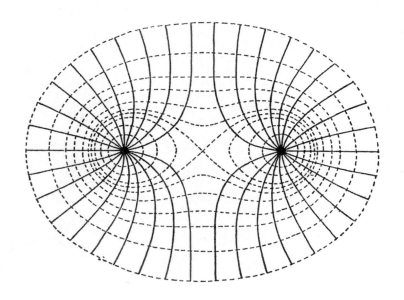

Fig. 89

In Fig. 89, the same set of curves is reproduced with their orthogonal trajectories (curved lines that everywhere cut the curves at right angles), which converge on one or other of the two foci. The full lines (the orthogonal trajectories) give a picture analogous to that which is obtained when one traces the "lines of force" of two like magnetic poles placed close together. This may be done by placing a piece of paper over two bar magnets with their north (or south) poles about 3 inches apart and sprinkling iron filings over the paper. On tapping the paper, the filings will arrange themselves in definite lines, that is, along the lines of force. Or if we consider two equal and similar electrical charges placed at the foci, then the orthogonal trajectories represent the lines of force due to these two charges, and the dotted curves represent the lines of equal potential. So we see here in the orthogonal trajectories a picture of two opposing forces clashing and repelling one another. Or again in the pointed egg-shapes of the ovals, we see that they have only just separated themselves from the unity of the "mother" figure and are straining towards one another; when they are pulled further apart they influence one another less and hence are merely rounded.

In his book *Growth and Form,* D'Arcy Thompson says that modern scientific investigations seem to lead to the general conclusion "that differences of electric potential play their part in the phenomena of cell division" (in the chapter, "On the Internal Form and Structure of the Cell"). He reproduces a drawing of the final stage in the first segmentation of the egg of Cerebratulus and side-by-side a diagram of a field of force with two similar poles; the similarity is striking.

Division

Just the same construction is used as for the pictures of the other processes. Two focal points are taken and a "constant quotient" (2 inches in Fig. 90a) is chosen, and then points are found, the quotient of whose distances from the two foci is always 2. Joining these points we obtain a pair of circles:

$$a/b = c/d = 2$$

When the foci are moved further apart, the form of the curve does not change; the circles simply become larger (Fig. 90b). Again, as in the case of subtraction, the commutative law is not valid (i.e., a/b does not equal b/a), and so the functions of the two foci are exchanged in constructing the second circle. It may be noted that two Apollonian circles of equal radius do not have the same relation to each other as have the two branches of the hyperbola (the curve of constant difference), which, of course, is a single curve.

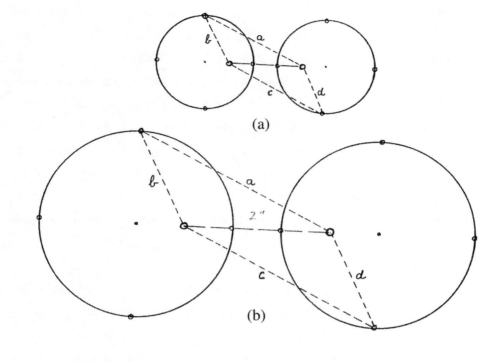

Figs. 90 a, b

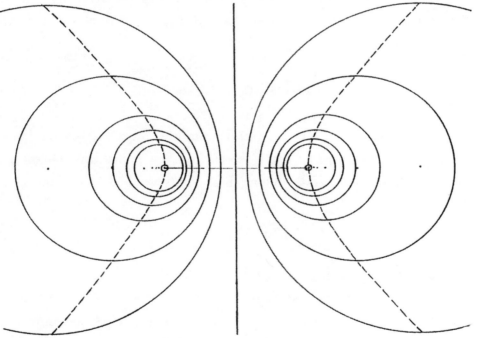

Fig. 91

If we keep the foci fixed but change the constant quotient, then we obtain a family of pairs of circles as in Fig. 91. As the constant quotient gets larger, the circles become smaller and farther apart, and their centers approach the foci closer and closer. The pairs of circles in the figure are the pictures of the constant quotients (going from outside to inside) 1⅔, 2, 3, 4, 5, and 6.

These circles, which are pictures of the process of division, are known as the circles of Apollonius. Apollonius of Perga (circa 225 B.C.) was one of the great mathematicians of Greece who wrote a systematic treatise on the conic sections, and it was he who gave these curves their names: ellipse, parabola, and hyperbola.

We have seen that the circles become smaller and smaller the greater the constant quotient, and their centers approach more and more closely to the foci. As the quotient approaches infinity, the circles tend to converge on the foci. The other limit for the drawing is when the quotient is unity; then the circles become infinitely big and touch one another. That is, this limit is the straight line that bisects at right angles the line joining the foci and the line at infinity of the plane.[11] It is interesting to notice that in passing from a constant quotient of 1⅔ to a constant quotient of 1—only a difference of $1/_2$—the geometrical picture passes from quite small circles to infinitely big ones.

If we join the highest and lowest points of the Apollonian circles by a smooth curve, we get a rectangular hyperbola (dotted line in Fig. 91), which is, as we have seen, a picture of the process of subtraction. Here we see a pictorial connection between the processes of division (circles of Apollonius) and of subtraction (hyperbola), and this we should expect since these two processes in arithmetic are closely related. We may say that division is a quick and short way of doing repeated subtraction. For example, suppose we ask the question: How many people can receive a gift of $5 out of a total sum of $20? We will work this out first by subtraction and then by division:

By subtraction:

$$
\begin{array}{rl}
\$20 & \\
-\ 5 & \text{1st subtraction} \\
\hline
15 & \\
-\ 5 & \text{2nd subtraction} \\
\hline
10 & \\
-\ 5 & \text{3rd subtraction} \\
\hline
5 & \\
-\ 5 & \text{4th subtraction} \\
\hline
0 &
\end{array}
$$

We see that we have to carry out the process of subtraction four times to use up all the $20. Thus four people can receive the gift of $5.

By division:

$$\$20 / \$5 = 4$$

A similar relationship between multiplication and addition (multiplication is a quick and short method of carrying out repeated addition) is shown pictorially by the circle drawn in the curves of Cassini picture (Fig. 88). The circumference of this circle passes through the highest and lowest points of the curves of Cassini within the flat-sided oval. The circle is a limiting case of an ellipse and is therefore a curve of addition; it is also related to the right angle (the angle in a semicircle is a right angle) corresponding to the hyperbola in Fig. 91, which is a rectangular hyperbola. It is also of interest to note that if we take the lemniscate as being typical of the "product" curves of Cassini, we have in its center part, where the two branches cut one another at right angles, the sign for multiplication (×). The sign for division (÷) can be seen in the "quotient" circles of Apollonius if the drawing of Fig. 91 is placed with the central line horizontal.

When the drawing of Fig. 91 is colored—especially if the smallest pair of circles is colored black—then one has the impression of a pair of eyes looking with a squint. When we observe an object acutely, we, as it were, "grip" the object with our eyes. The axes of our eyes converge onto the object and "hold it" in vision. Thus, when we look at an object, especially when it is close to us, the pupils of the eyes are inclined inwards, that is, the eyes have a slight squint. Moreover when we observe anything acutely and with concentration, we may say that we are within a process of calculation. We even speak of the "calculating look or glance" of a human being. Now which arithmetical process is constantly used by the eyes in observing the objects of the outer world? It is surely the process of division, for the eyes divide the one object into two pictures, one on each retina, which then become again united into a single image by an elementary act of thought, by the activity of the human ego. The genius of language reveals to us the same relationship: the Latin *dividere,* to divide, is connected with *videre*, to see; in English the word *vision* is contained in the word *division.*

It may be noted that the orthogonal trajectories of the Apollonian circles resemble the curves of the lines of force of two unlike magnetic poles or electric charges. This resemblance becomes closer the further the poles are apart.

In giving such drawings to children, one is not giving them something arbitrary, but something with a real, deep content. The connections with outer nature and with man himself that the teacher indicates are not fanciful but fundamental realities. So the children will gain their knowledge of geometry on a sound and sure basis, and it will become increasingly living and vital for them when, in their later school years, they come to have a more intellectual understanding as the complement of what they have been experiencing imaginatively. By such a method of teaching, the conception that "God is eternally geometrizing" becomes more and more of a reality.

Chapter 6

The Five Regular Solids

The five regular solids, as their name implies, are solid figures having identical regular faces, equal edges, and equal angles. They are sometimes known as "Platonic bodies" because of their significance in the molecular theory and cosmogony propounded by Plato in his *Timaeus*. These five polyhedra are show drawn in perspective in Fig. 92. Their names are (a) cube, (b) octahedron (8 faces), (c) tetrahedron (4 faces), (d) icosahedron (20 faces), and (e) dodecahedron (12 faces). They were first studied in the school of Pythagoras, who may have brought his knowledge of the first three from Egypt. The Pythagoreans related the tetrahedron to the element of fire, the octahedron to air, the icosahedron to water, the cube to earth, and the dodecahedron to the universe. They also knew that all five polyhedra can be inscribed in a sphere, which they considered the most perfect of all solid bodies. The study of these figures was continued in the school of Plato (c. 380 B.C.). The dodecahedron undoubtedly had a mystical or religious significance; a number of bronze models of ancient Celtic origin are still to be seen in various museums, and a stone dodecahedron found in northern Italy belongs to a prehistoric period. In the Middle Ages, astrologers and astronomers occupied themselves with these regular solids. Johannes Kepler made a considerable study of them, and in 1596, he believed that he had found a relationship between them and the number and distances of the planets. He set forth his discovery in his *Mysterium Cosmographicum (The Mystery of the Universe)*; he began by constructing a series of regular polygons of such size that a circle could be inscribed in each and at the same time should be the circumscribed circle to the next member of the series. He hoped that the radii of successive circles might be proportional to the distances of successive planets, but in this he was disappointed. This, however, led him to apply the same method to the regular polyhedra; he calculated the radii of the pairs of spheres that can be inscribed and circumscribed to the five regular solids placed one within the other, the sun being at the center of this system. The result satisfied him that he had discovered a fundamental secret of the universe: The radii of the inscribed and circumscribed spheres of an octahedron were fairly closely proportional to the greatest distance of Mercury and the least distance of Venus, respectively, from the sun. The radii of the inscribed and circumscribed spheres of an icosahedron were found to correspond to the greatest distance of Venus and the least distance of the Earth. The dodecahedron, tetrahedron, and cube could similarly be placed between the successive orbits of the Earth and the three outer planets, Mars, Jupiter, and Saturn. Kepler realized that the relationship was not perfect, but at that time, he attributed the discrepancies to inaccurate observations. His book, *Mysterium Cosmographicum,* was the means of introducing him to the Danish astronomer Tycho Brahe, and from that moment Kepler's further researches, based on Tycho's observations, which he trusted completely, led him away from his ingenious theory and towards his epoch-making discovery of the elliptical orbits of planetary motion.

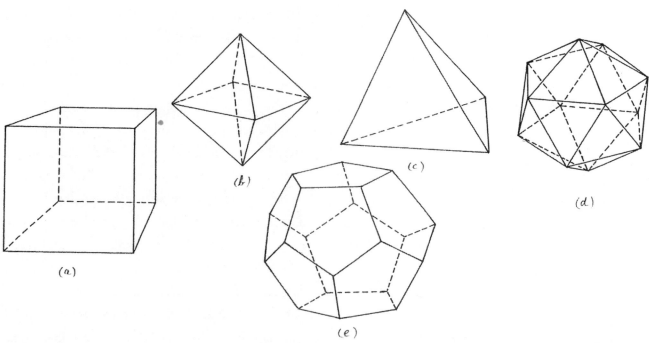

Fig. 92

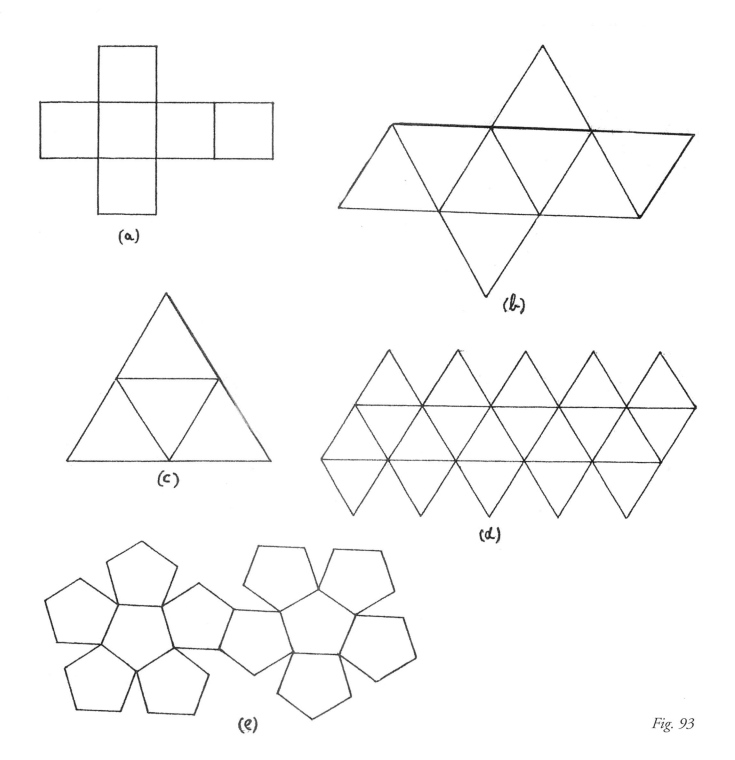

Fig. 93

One of the chief values to be gained in studying the five regular solids with children of, say, 15 years of age, is in their making models of these figures. Such an activity calls forth a high degree of accuracy and quite skillful manipulation. It was the great German artist Albrecht Dürer (1471–1528) who first showed how to construct the regular solids by drawing the surface polygons all on one piece of paper or thin cardboard and then folding along the connected edges. These "nets," as they are now called, are shown in Fig. 93, and they form, when folded and the edges are stuck together (flanges not shown), the corresponding solids of Fig. 92.

It may quite easily be proved that there cannot be more than five regular polyhedra:

At least three plane angles are required to form a solid angle, and the sum of such plane angles must be less than 360°, otherwise the plane angles would all lie in the same plane and there would be no solid angle. It follows that each angle of the faces forming a regular polyhedron must be less than 120°. That is, the faces can only be equilateral triangles, squares, or pentagons; for the angle of a regular hexagon is 120°, and any regular polygon of more than six sides has an angle greater than 120°. Let Q represent the number of degrees in a face-angle. When the faces are equilateral triangles, $Q = 60°$. Then (*i*) $3Q = 180°$, (*ii*) $4Q = 240°$, (*iii*) $5Q = 300°$, ($6Q = 360°$). Thus three, four, or five equilateral triangles, and not more than five, can be used to form a solid angle in a regular polyhedron. When the faces are squares, $Q = 90°$. Then (*iv*) $3Q = 270°$ ($4Q = 360°$). Thus three squares, and only three, can be used to form a solid angle in a regular polyhedron. When the faces are pentagons, $Q = 108°$. Then (*v*) $3Q = 324°$ ($4Q = 432°$). Thus three regular pentagons, and only three, can be used to form a solid angle in a regular polyhedron.

Therefore there can only be five regular polyhedra.

The nets shown in Fig. 93 are, of course, an illustration of the above proof. For we see in Fig 93c, b, d, three, four, and five equilateral triangles, respectively; in Fig. 93a, three squares; and in Fig. 93e, three regular pentagons around an angular point, that is, around a point that will become an angular point of the solid figure.

The drawings of Fig. 94 are plane views of the five Platonic solids inscribed in circles of the same radius. In these views, all the angular points, all the edges, and all the faces of each solid can be seen, assuming the figures are made of glass. Other positions would give plane views in which points, edges, and surfaces would be obscured, for example, a square being the plane view of a cube. It should be noted that there are two plane views of the tetrahedron (Figure 94c), which show all its points, edges, and faces, while there is only one such view of each of the other solids. Table 6-1 gives the numbers of angular points, edges, and faces for each of the five solids.

Table 6-1

Name of Polyhedron	Number of Angular Points	Number of Edges	Number of Faces
Tetrahedron	4	6 (3 per angle)	4 equilateral triangles (3 per angle)
Octahedron	6	12 (4 per angle)	8 equilateral triangles (4 per angle)
Cube	8	12 (3 per angle)	6 squares (3 per angle)
Icosahedron	12	30 (5 per angle)	20 equilateral triangle (5 per angle)
Dodecahedron	20	30 (3 per angle)	12 regular pentagons (3 per angle)

Euler discovered a formula relating the numbers of angular points (p), edges (e), and faces (f):

$$p + f = e + 2$$

This formula applies only to the five Platonic solids.

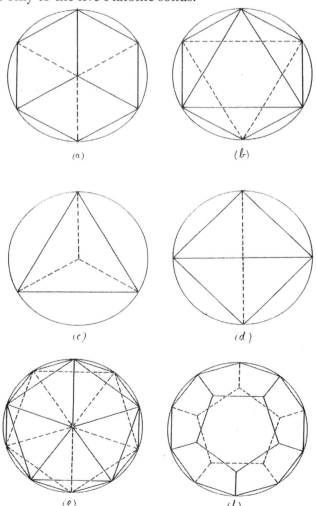

Fig. 94

A very important structural relationship is to be found among the solids themselves. The cube and the octahedron are intimately related to one another in that the one may be transformed into the other: If the eight corners of a cube are truncated symmetrically, eight equilateral triangular faces appear until, by further cutting, the octahedron arises. If the six corners of an octahedron are truncated symmetrically, six square faces appear until by further cutting the cube arises. The cube and the octahedron are said to be "dual" figures because the number and configuration of the faces of the one is equal to the number and configuration of the corners of the other and vice versa. By such a reciprocal transformation, the icosahedron and dodecahedron are seen to be dual figures, while the tetrahedron is self-dual. That is, it becomes transformed into itself. Fig. 95 shows an octahedron inscribed in a cube and illustrates the dual nature of these two figures. A sphere is also shown inscribed in the cube and therefore circumscribed to the octahedron. Thus, the six planes of the cube are six tangent planes of the sphere, and the points of contact of these planes with the sphere form the angular points or corners of the inscribed octahedron. This is a fine example of how in regular spatial forms, point corresponds to plane and plane to point. These two forms are in every respect polar opposites with regard to point and plane. Thus each corner of the cube is in polar correspondence with a plane of the octahedron, and each corner of the octahedron is in polar correspondence with a plane of the cube with reference to the sphere.

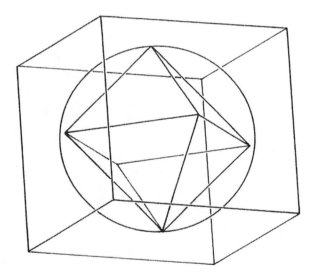

Fig 95

There are other structural relationships that arise between certain pairs of the solids. For example, a tetrahedron is contained by a cube, the angular points of the tetrahedron being four of the eight angular points of the cube, and its edges being diagonals of the six faces of the cube (Fig. 96). The other four corners of the cube and the other six diagonals of its faces give a second tetrahedron. Or again, the midpoints of the six sides of a tetrahedron are the angular points of an octahedron (Fig. 97). A cube may be circumscribed round an icosahedron, all the 12 angular points of the latter lying on the faces of the cube. If all the edges of a dodecahedron are produced, their points of intersection are the 12 angular points of an icosahedron. And if all the edges of an icosahedron are produced, their points of intersection are the 20 angular points of a dodecahedron.

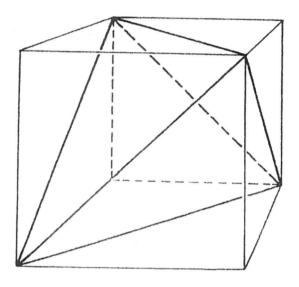

Fig. 96

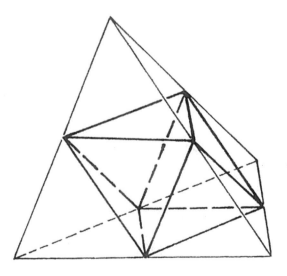

Fig 97

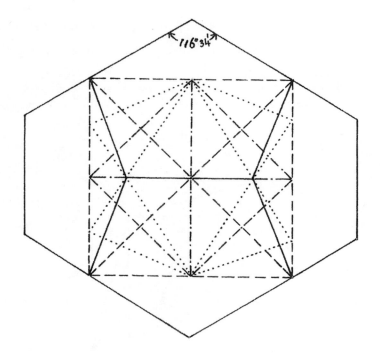

Fig. 98

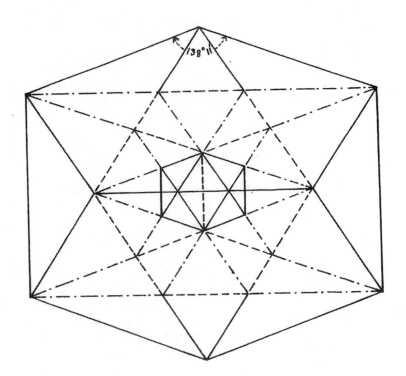

Fig. 99

Fig. 98 is the orthogonal projection of all five Platonic bodies in one figure. (Pans and elevations are orthogonal projections—the projecting lines are at right angles to the plane of projection). The full line figure—the largest—is the dodecahedron. Within this is the cube (the broken line square) and the tetrahedron (broken line square with diagonals). Within the cube are the octahedron and the icosahedron. (The octahedron is drawn with "chain" lines; the horizontal chain line coincides in part with the horizontal edge of the dodecahedron. The icosahedron is drawn with dotted lines; the horizontal edge coincides with that of the dodecahedron and the vertical edges coincide with parts of the edges of the cube.) It should be noted for purposes of construction that the dihedral angle[12] of a dodecahedron is 116° 34' and that the outside edge of the dodecahedron projection is contained in a square. This arises from the fact that 12 of the angular points (out of 20) of the dodecahedron lie on the surfaces of a cube, 2 on each of the 6 faces.

Fig. 99 is an orthogonal projection showing the reciprocal transformation of an icosahedron into a dodecahedron by producing the edges as referred to above. The edges of the small icosahedron (shown in solid line) are produced to give the angular points of a dodecahedron (shown in broken line), and the edges of this figure are produced to give rise to a large icosahedron (shown in solid line except where edges correspond with those of the dodecahedron; the producing of the edges is shown by chain line). Again for purposes of construction, the dihedral angle of an icosahedron is 138° 1', and the outside edge of the icosahedron projection is contained in a square. This arises from the fact that the 12 angular points of an icosahedron all lie on the surface of a cube, 2 on each of the 6 faces.

Reference to Tables 6-2 and 6-3 shows that in all the numerical values of length of edge, surface area, and volume of both the icosahedron and the dodecahedron, the golden number appears. Also, interesting proportions are to be found between the different figures. For example, from Table 6-2, where the calculations are made assuming a circumscribing sphere of unit radius, we see that

$$\frac{\text{Edge of cube}}{\text{Edge of dodecahedron}} = G$$

while from Table 6-3 where the sphere to which the edges of the solids are tangents is considered of unit radius, we find that

$$\frac{\text{Edge of icosahedron}}{\text{Edge of dodecahedron}} = G \qquad \frac{\text{Edge of octahedron}}{\text{Edge of icosahedron}} = G$$

while the edge of the tetrahedron is double that of the cube, and the surface areas of the octahedron and tetrahedron are equal.

All four other regular solids may be inscribed in a cube. Fig. 100 is an orthogonal projection of an icosahedron and a dodecahedron inscribed in a cube in such a manner that the 12 angular points and 6 of the edges of the dodecahedron lie on the surface of the cube. We then find geometrically that

$$\frac{\text{Edge of icosahedron}}{\text{Edge of dodecahedron}} = G$$

which we have also found by calculation (see Table 6-3). The remaining 8 angular points of the dodecahedron are coincident with the 8 angular points of a smaller cube whose edge is equal to the edge of the icosahedron. The sides of the 2 cubes are in golden ratio. (The orthogonal projection of the icosahedron is shown in chain line, the dodecahedron in solid line, and the large and small cubes and construction lines in broken line.) It will be noticed that the figure contains a number of golden rectangles (see Chapter 4).

The fact that there are only five regular solids, that they are related among themselves in such an astonishing variety of ways, and that their mathematical proportions, including the golden ratio, are so significant leads us to realize that they are fundamental to an understanding of space. As forms in nature, they are readily to be found, although some of them only in microscopic organisms. The first three solids occur in the crystalline structure of a number of common minerals: copper antimony sulfide (tetrahedrite) and zinc sulfide (zinc blend) in tetrahedra; lead sulfide (galena), rock salt, and fluorite or fluorspar in cubes (fluorspar also occurs in octahedra); and magnetic oxide of iron (magnetite) in octahedra. The icosahedron and dodecahedron never occur in the mineral kingdom because they have a fivefold axis of symmetry, and only two-, three-, four-, and sixfold axes of symmetry are possible in crystallography. This is a fundamental law concerned with the whole theory of the partitioning of space. In the organic world, the cube is found, at least in outline, in the skeleton of the hexaotinellid sponges; the tetrahedron, octahedron, icosahedron, and dodecahedron in the various forms of radiolarian. These minute unicellular marine organisms form the bed of the deep oceans; their skeletons are mostly of silica. Haeckel, in his *Monograph of the Challenger Radiolaria* (1887), estimated the number of known forms at 4,314 species, included in 739 genera. Fig. 101 shows the skeletons of various radiolarians, after Haeckel: (a) *Circoporus octahedrus,* (b) *Circogonia icosahedra,* (c*) Circorrhegma dodecahedra,* (d) *Circospathis novena* (D'Arcy Thompson, *On Growth and Form*). It may be noted that the faces, which are not necessarily planar, are formed of a minute hexagonal network of silica.

Four more solid figures are, technically speaking, regular. They differ from the five platonic solids in that they have re-entrant angles. The small and great stellated dodecahedra were discovered by Kepler (1571–1630) and the great icosahedron and the great dodecahedron by the French mathematician Louis Poinsot (1777–1859). They are very beautiful figures but their construction from nets is more difficult (see Cundy and Rollett, *Mathematical Models*). The plan views of the two Kepler solids are shown in Fig. 102 (the small stellated dodecahedron) and Fig. 103 (the great stellated dodecahedron). We see how these orthogonal projections are related to the pentagram and the decagon star, and as we should expect, the golden ratio appears again. Thus,

$$\frac{OB}{OC} = \frac{OA}{OB} = G \text{ (both figures)}$$

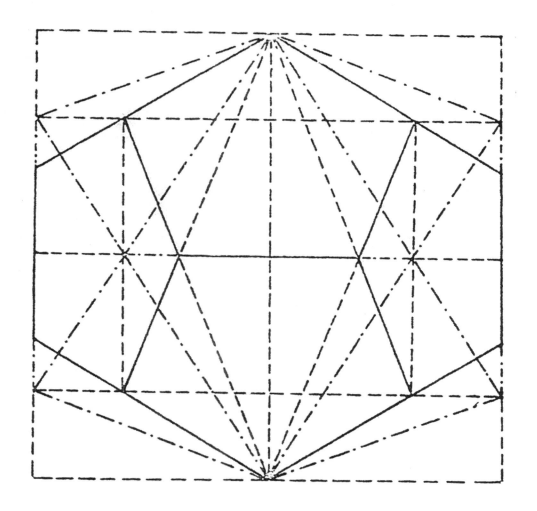

Fig. 100

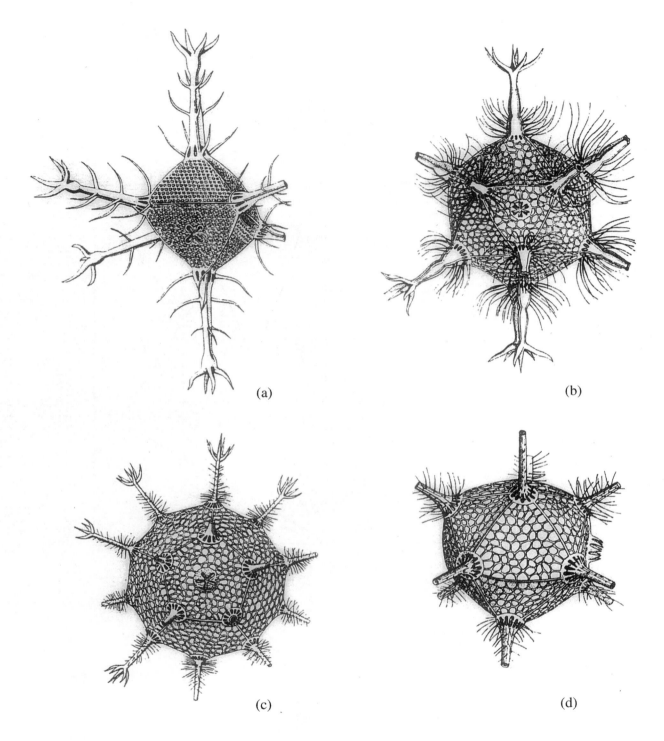

Fig. 101

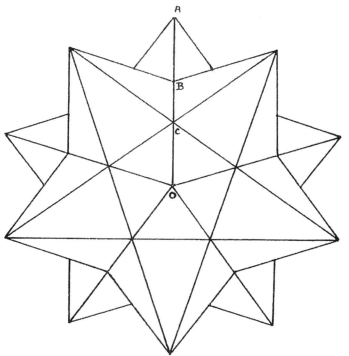

Fig. 102

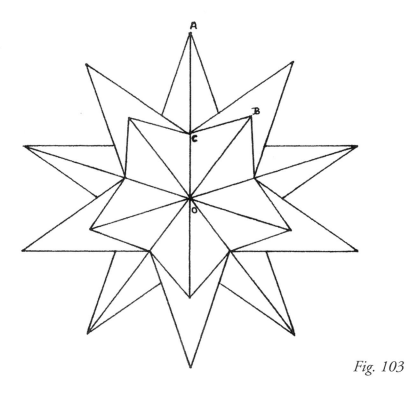

Fig. 103

137

Table 6-2

	Radius of Circumscribed Sphere	Length of Edge	Surface Area	Volume
Cube	1	$\dfrac{2}{\sqrt{3}}$ (= 1.155)	8	$\dfrac{8}{3\sqrt{3}}$
Octahedron	1	$\sqrt{2}$ (= 1.414)	$4\sqrt{3}$	$\dfrac{4}{3}$
Tetrahedron	1	$\dfrac{4}{\sqrt{6}}$ (=1.633)	$\dfrac{8}{\sqrt{3}}$	$\dfrac{8}{9\sqrt{3}}$
Icosahedron	1	$\dfrac{4}{\sqrt{10+2\sqrt{5}}}$ (= 1.1051) $= \dfrac{2}{\sqrt{G_2+1}}$	$\dfrac{40(3)}{5+\sqrt{5}}$ $= \dfrac{20\sqrt{3}}{G_2+1}$	$\dfrac{2\sqrt{10+2\sqrt{5}}}{3}$ $= \dfrac{4\sqrt{G^2+1}}{3}$
Dodecahedron	1	$\dfrac{\sqrt{5}-1}{\sqrt{3}}$ (=0.714) $\dfrac{2}{G\sqrt{3}}$	$\sqrt{5}\,(\sqrt{5}-1)\,\sqrt{10+2\sqrt{5}}$ $= \dfrac{4\sqrt{5}}{G}\sqrt{G^2+1}$	$\dfrac{2\sqrt{5}\,(\sqrt{5}+1)}{3\sqrt{3}}$ $\dfrac{4G\sqrt{5}}{3\sqrt{3}}$

Table 6-3

	Radius of Sphere to Which Edges Are Tangent	Length of Edge	Surface Area	Volume
Cube	1	$\sqrt{2}$ ($=1.414$)	12	$2\sqrt{2}$
Octahedron	1	2	$8\sqrt{3}$	$\dfrac{8\sqrt{2}}{3}$
Tetrahedron	1	$2\sqrt{2}$	$8\sqrt{3}$	$\dfrac{8}{3}$
Icosahedron	1	$\sqrt{5}-1$ ($=1.236$)	$10\sqrt{3}\,(3\sqrt{5})$ $=\dfrac{20\,(\sqrt{3})}{G^2}$	$\dfrac{10\,(\sqrt{5}-1)}{3}$ $=\dfrac{4\sqrt{5}}{3G}$
Dodecahedron	1	$3-\sqrt{5}$ $=\dfrac{2}{G^2}$	$6\sqrt{10(25-11\sqrt{5})}$ $=\dfrac{12\sqrt{5}\sqrt{G^2+1}}{G^3}$	$=2(3\sqrt{5}-5)$ $=\dfrac{4\sqrt{5}}{G^3}$

Note: $G = \dfrac{\sqrt{5}+1}{2}$ and $\dfrac{1}{G} = \dfrac{\sqrt{5}-1}{2}$ (G is the golden number.)

Chapter 7

The Conic Sections

The study of the important curves of the conic sections—ellipse, parabola, and hyperbola—is undertaken in most schools today only with pupils in the higher classes. A knowledge of these curves is required for advanced level examinations, and then they are usually treated only from an analytical point of view. An older boy or girl will know the equations of the curves and some of their elementary analytical properties but will have little or no realization of their real nature and essence, namely that they are fundamentally one and the same curve in three different metamorphoses and that they may all be considered as curves of addition.[13] Treated in a pictorial and imaginative way, the conic section curves can be studied by children of 15 or 16 years, and they will then gain a firm basis of knowledge and understanding for the later more abstract and intellectual approach. They will also realize—which they will never do from the purely analytical aspect—that these curves are fundamental to an understanding of form in the universe, Earth, and man. Here again, geometry becomes a real cultural subject and not merely a subject that they must learn in order to pass their examinations.

As the name implies, a conic section is the intersection of a plane with a cone. The great mathematicians of the Golden Age of Greek mathematics were the first to study in detail the curves of the conic sections, and this work seems to have originally issued from Plato's Academy with its inscription, "Let none unversed in geometry enter." Menaechmus (c. 375–325 B.C.), an associate of Plato, acquired a great reputation as a teacher of geometry and was appointed one of the tutors of Alexander the Great, who once asked him to make his proofs shorter. Menaechmus replied: "Though in the country there are private and even royal roads, yet in geometry there is only one road for all." He discussed the conic sections, which for long afterwards were known as the "triads of Menaechmus."

During the next hundred years, great progress was made in the study of these curves by a number of Greek mathematicians, including Euclid and Archimedes, although the writings of these two are lost. The crowning work in this subject was carried out by Apollonius of Perga (c. 260 B.C.–200 B.C.), known as the Great Geometer, who was the author of eight books on conics, the first four of which have come down to us in Greek and the next three in Arabic, the last one being lost. (We have already referred to Apollonius in Chapter 5.) Greek mathematics culminated in the work of Apollonius, and we may say that the Greeks studied these curves on their own account because they recognized that they were fundamental to an understanding of form everywhere.

For the next eighteen centuries, little progress was made in further knowledge of conics, and indeed they seem to have been almost forgotten until the work of the astronomer Johann Kepler (1571–1630) revived interest in them. Kepler placed the Sun in a world focus, with the planets revolving round it in elliptical orbits. Up to this time, the treatment of the conic sections had been entirely descriptive and synthetic, but with Descartes (1596–1650), a new treatment became possible through his algebraization of geometry—what we today call analytic or coordinate geometry—and it is this method that is chiefly used today in the study of these curves. However, contemporary with Descartes were two other great French mathematicians, Gerard Desargues (1593–1662) and Blaise Pascal (1623–1662), who introduced a new method of investigating the subject now known as projective geometry. This has made steady progress during the past four centuries and during the twentieth and twenty-first centuries has increasingly come to be recognized as a geometry of the first importance. The following chapter of this book is devoted to the teaching of projective geometry, which involves a further consideration of the conic section curves beyond the scope of the present chapter.

We will now consider these curves—ellipse, parabola, and hyperbola—from a descriptive, pictorial point of view. The children will of course be familiar with the form of a cone, but they must now realize that the geometrical conception involves a "double cone" as they know it. The following drawing (Fig. 104) is an elevation view of such a cone. The dotted lines indicate the cutting planes, and against these lines is written the name of the curve formed by the plane cutting the cone. It is a good exercise in imagination to picture the form of the curve obtained as the cutting plane rotates through a right angle. It comes as somewhat of a surprise that the cutting of a cone in a certain direction gives the same curve (an ellipse) as does the cutting plane of a cylinder.

It is clear from this drawing that during the rotation of the cutting plane through a right angle (the plane passing through a point P on the axis), there is only one circle (cutting plane at right angles to the axis), only one parabola (cutting plane parallel to the slant edge of the cone), whereas there are many ellipses and hyperbolas.

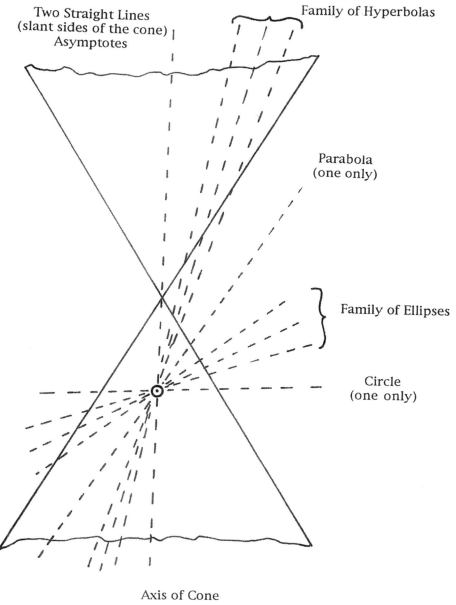

Fig 104

Although it takes rather a long time and involves somewhat complicated drawings, the making of models of cones (or rather half-cones) cut by a plane in these three sections is an excellent exercise in accurate drawing and construction for children of this age. Such models also provide the confirmation of what they have imagined as the conic section curves, from their consideration of the drawing of Fig. 104. The following three sets of drawings, Figs. 105, 106, and 107, show the constructions necessary for making half-cones out of thin cardboard cut in their sections of ellipse, parabola, and hyperbola. The notes and lettering on the drawings will enable the reader to follow the construction. The half-cone in each case has the same dimensions, and therefore the notes for the first set of drawings apply to the other two. The cone may of course be varied as regards shape and size. (It should be pointed out that in the drawings, no means are shown for sticking the various cardboard parts together. For this purpose flanges must be added, and except for the long flange along Oa, these should be kept small as they are on curves. It may be advisable not to add flanges to the circular base and the curved section, but to stick them into position with small independent flanges.) It is not easy to get a good fit in making these models even if the drawings are done very accurately, because it is almost impossible to make allowance for the thickness of the cardboard.

Notes on Models of Cones Cut in Their Sections
(Figs. 105, 106, 107)

Circumference of base of cone = $\pi \times$ diameter = $\frac{22}{7} \times 3\frac{1}{2}$ inches = 11 inches (full scale).

This circumference, when the slant surface of the cone is "rolled out" flat, becomes an arc of a circle whose radius is the length of the slant side = 4.4 inches.

Therefore, the circumference of the circle of radius 4.4 inches = $2 \times \frac{22}{7} \times 4.4$ inches

$= 27\frac{2}{3}$ inches (approx.)

Therefore, the angle of segment of this circle = $\frac{11 \times 3}{83} \times 360°$

$= 143°$ (approx.).

The segment of the circle with the angle of 143° is therefore the "flattened out," slant side of the cone. Cut along the curve drawn on this segment, and when the

corresponding parts of the two radii $\left\{\begin{array}{l}\text{Oa}\\ \text{Oa}\\ \text{Oe}\end{array}\right.$ and $\left\{\begin{array}{l}\text{Oa}\\ \text{Oa}\\ \text{Oe}\end{array}\right.$

are joined together by a flange, this curve will form the $\begin{cases} \text{ellipse,} \\ \text{parabola,} \\ \text{hyperbola,} \end{cases}$

which the cutting of the cone has produced.

The right-angled triangle constructional drawing gives the lengths necessary for plotting the points required for drawing the curve on the flattened out, slant side. This triangle is half the elevation of the cone. The lettered lengths explain the construction. (The above notes apply to all three cases.)

It is thus clear from the outset that these three curves "belong together" and must be closely related to one another since they are all obtained from the cutting of a cone. There are a number of constructions for drawing the conic section curves, and we shall choose to begin with one that has a certain consistency in it as applied to all three curves. Later on in the chapter, other constructions will be shown.

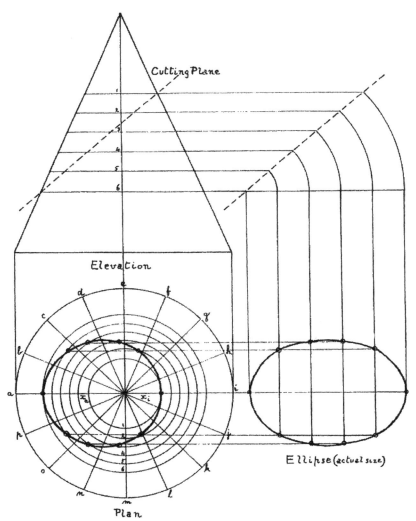

Fig. 105 a

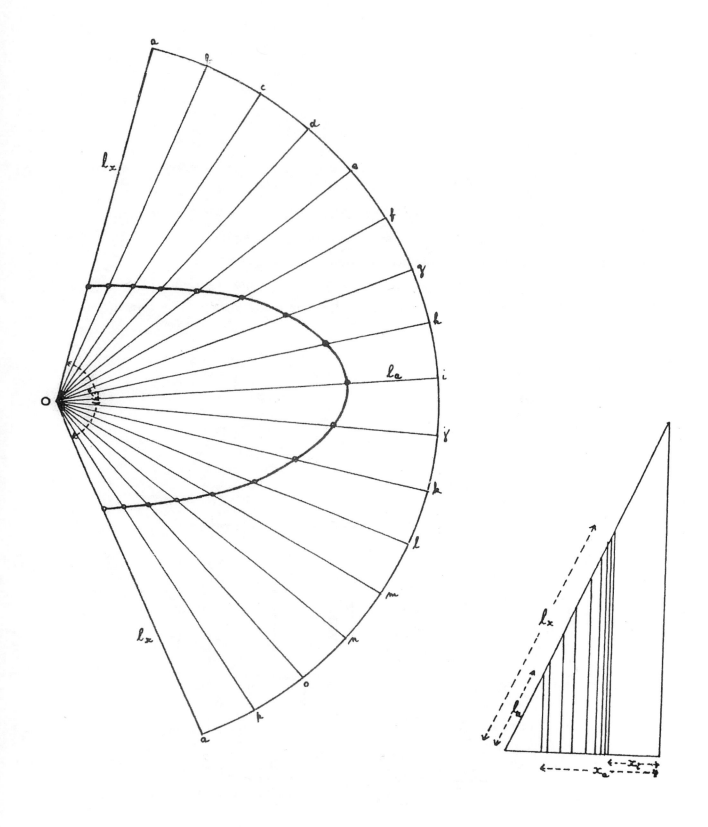

Fig. 105 b

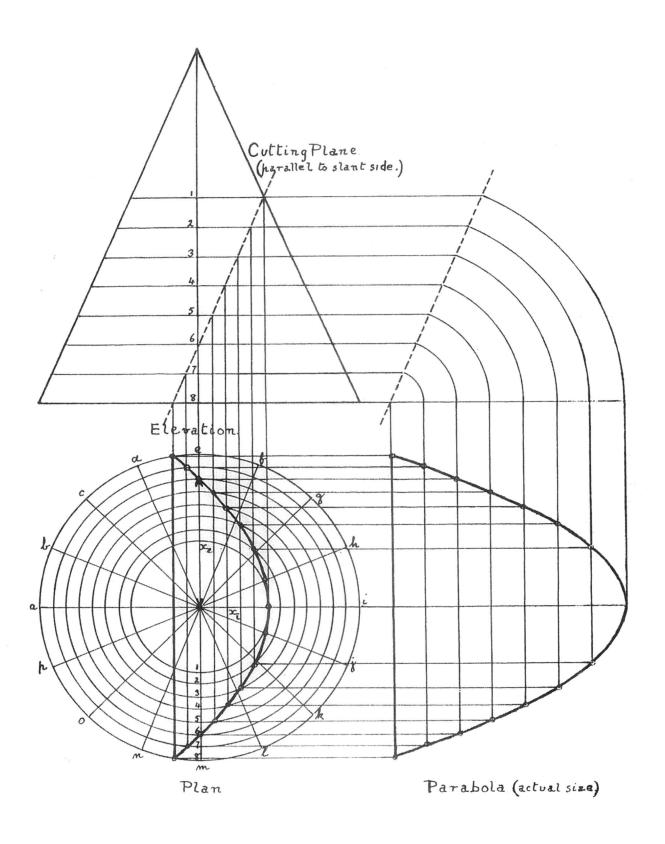

Fig. 106 a

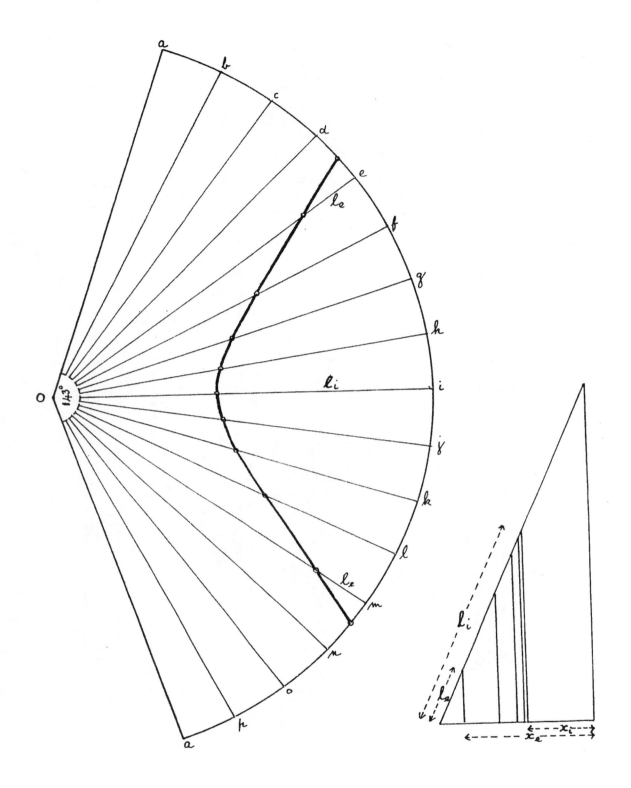

Fig. 106 b

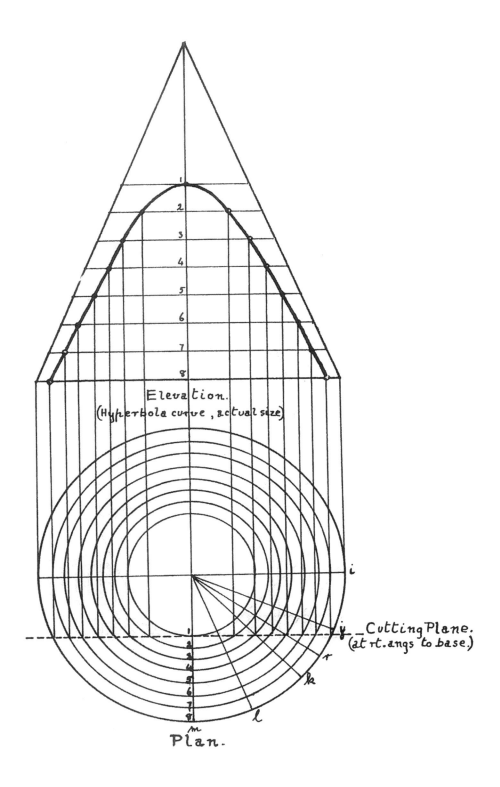

Fig. 107 a

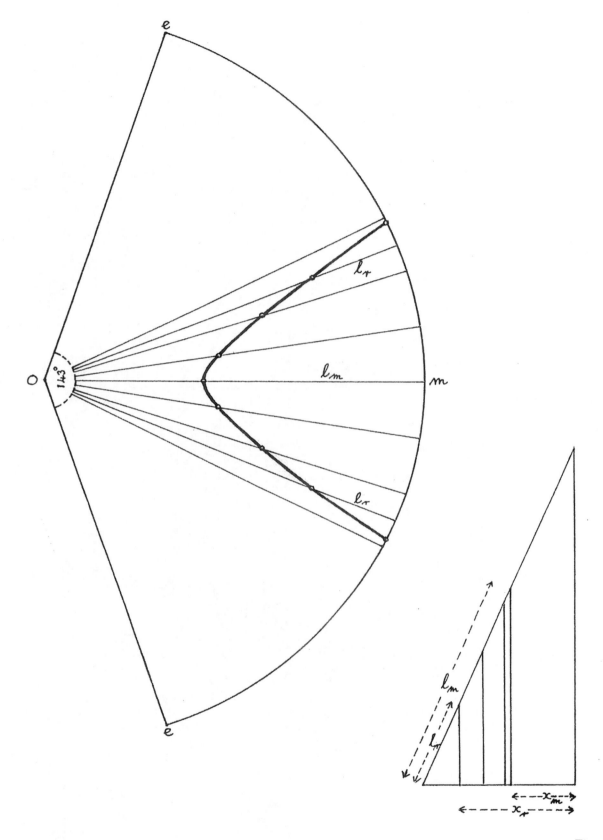

Fig. 107 b

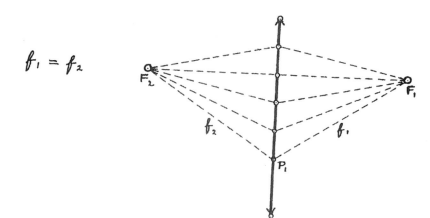

Fig. 108

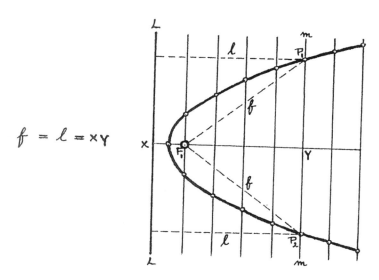

Fig. 109

Fig. 110

151

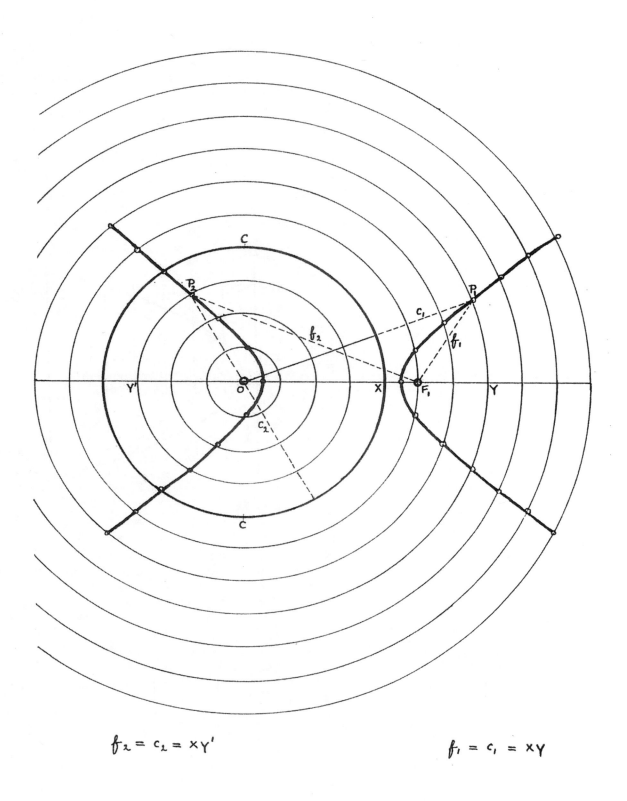

$f_2 = c_2 = XY'$ $f_1 = c_1 = XY$

Fig. 111

Let us consider the symmetry of the human form as between right and left: If a man stands with his arms outstretched horizontally, then between fingertip and fingertip, there is an "axis of symmetry" that exactly divides left from right (see Fig. 108). Thus we see that the axis of symmetry between two points is a straight line, which is the right bisector of the line joining the two points. Now we will change one of the points (say F_2) into a straight line and ask: What is the axis of symmetry between a point and a line? Our problem now is to find all the points in the plane of the paper that are equidistant from the given point F_1 and the given line L and join them together by a smooth curve (see Fig. 109). To do this, we draw a family of straight lines parallel to the given line L. Consider one of these lines, say, mm. Then somewhere along it, there must be two points, each equidistant from the fixed point F_1 and the fixed straight line L, one above and one below the line through F_1 at right angles to line L. Obviously, since mm is parallel to L, the distance of these points from the line L is XY. Thus with center F_1 and radius XY, make arcs cutting mm in P_1 and P_2; then P_1 and P_2 are the required two points. Repeat this construction for points on the other parallel lines, draw a smooth curve through all the points so obtained, and the required line of symmetry is obtained. We see that it is a parabola. Thus a parabola is the line of symmetry between a point and a line. The straight line L is now bent into a circle; this may be done so as to enclose the point F_1 or so as to leave the point F_1 outside the circle. The first case of the circle C enclosing the point F_1 is shown in Fig. 110, and we see that the line of symmetry between a point and a circle enclosing it is an ellipse. The construction is the same as for the parabola except that the points through which the curve is drawn are found not on parallel lines but on concentric or parallel circles. Now for the second case where the circle is formed outside the point, we find that the line of symmetry between such a point and a circle is an hyperbola (Fig. 111). It will be noticed that all distances of points from the circle lie along "radius lines" passing through its center, that is, along normals to the circle, and that these distances giving the points lying on the left-hand "wing" of the hyperbola are the longer of the two possible distances. For example, the point P_2 has two distances from the circle lying along the diameter through P_2; it is the longer one (XY^1) that we use in determining this point. In the foregoing constructions of the ellipse and the hyperbola in which a circle is used, we see that the center O of the circle gives a second point corresponding to the point F_1; in the future, we shall call this second point F_2. These two points F_1 and F_2 are called the foci of the curves. In the case of the parabola, there appears to be only one focus, F_1. This will be discussed later in the chapter.

The drawings that now follow (Figs. 112–120) show families of conic section curves obtained by the above method of construction.

> Fig. 112   A family of parabolas having the same focus but the straight line moving rhythmically. Note that the points determining the curves are formed by concentric circles (with focus as center) cutting the parallel lines, these lines being tangents to the circle.

Fig. 113   A family of parabolas constructed with reference to the same straight line but the focus moving rhythmically.

Fig. 114   One family of confocal parabolas forming the orthogonal trajectories of the another family.

Fig. 115   A family of ellipses having the same foci but the circles moving rhythmically. Note that the points determining the curves are formed by the intersection of two series of concentric circles (with foci as centers) that touch one another in pairs.

Fig. 116   A family of ellipses constructed with reference to the same circle but the focus moving rhythmically. Note that the second focus of each ellipse is the center of the circle.

Fig. 117   A family of similar confocal ellipses. In each ellipse,
$$\frac{\text{Distance between foci}}{\text{Major axis}} = \frac{2}{3}$$

Fig. 118   A family of ellipses constructed with reference to the same circle but the focus moving along the arc of a circle.

Fig. 119   A family of hyperbolas having the same foci but the circle moving rhythmically. Note that the points determining the curves are formed by the intersection of two series of concentric circles (with foci as centers) that touch one another in pairs.

Fig. 120   A family of hyperbolas constructed with reference to the same circle but the focus moving rhythmically. Note that the second focus of each hyperbola is the center of the circle.

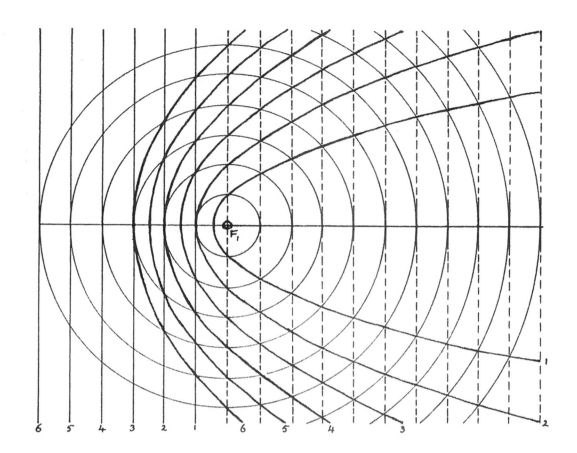

Fig. 112

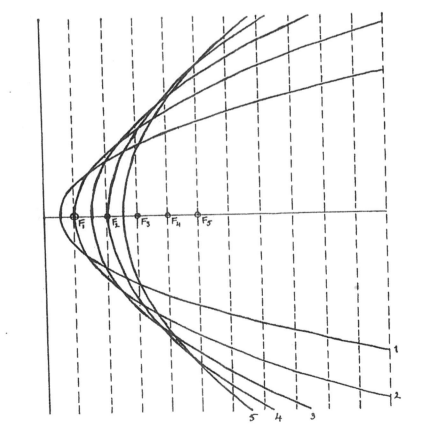

Fig. 113

155

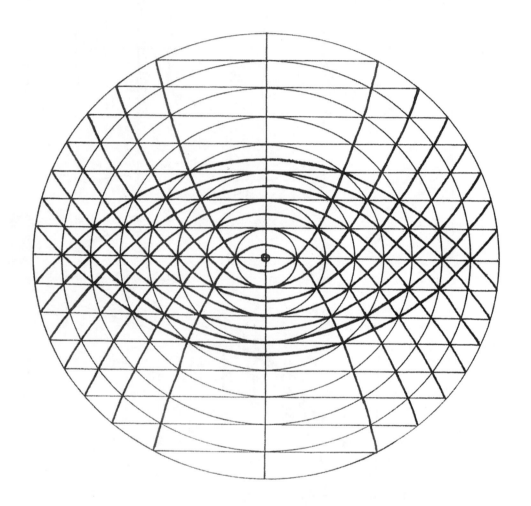

Fig. 114

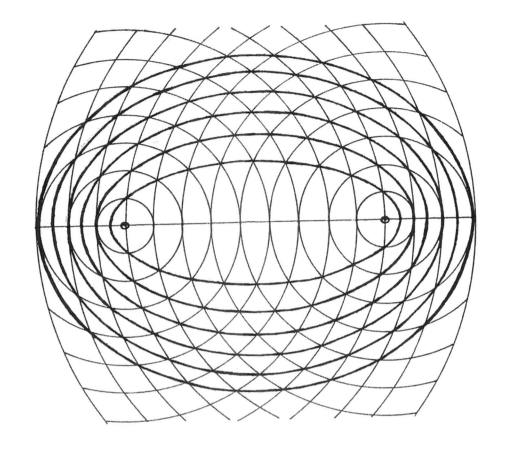

Fig. 115

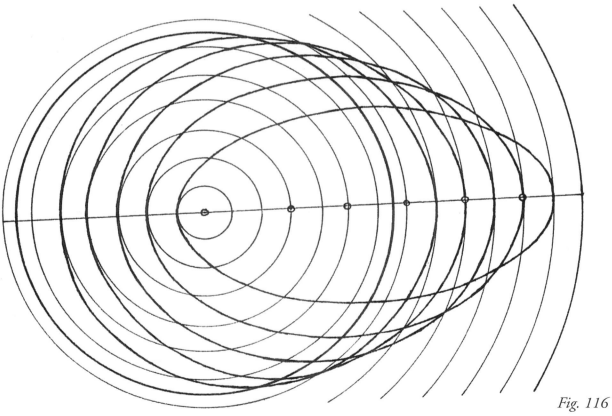

Fig. 116

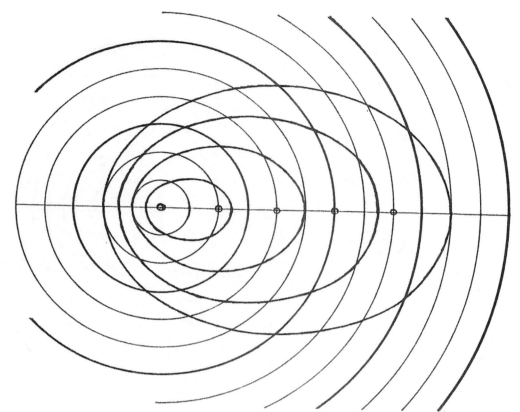

Fig. 117

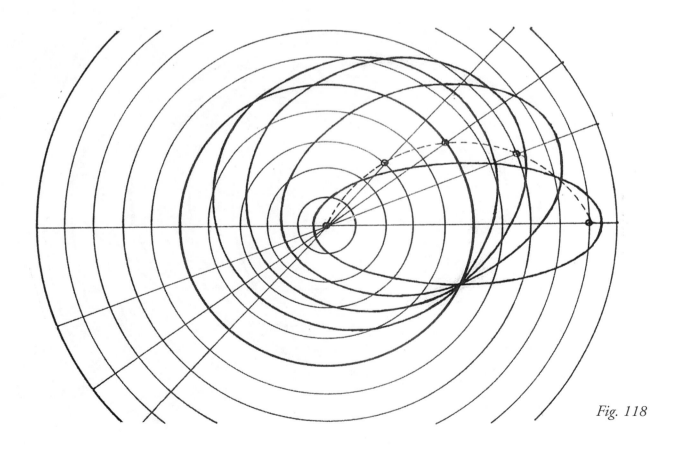

Fig. 118

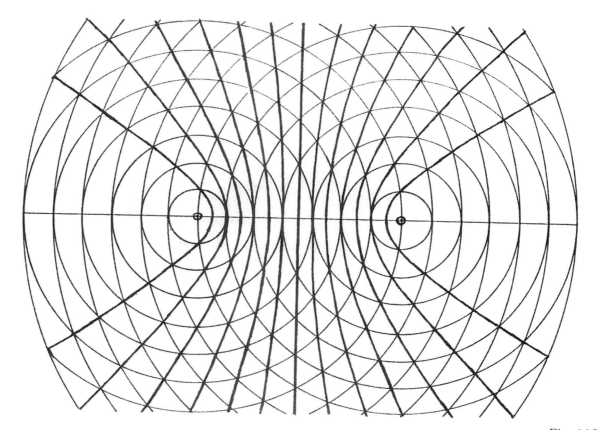

Fig. 119

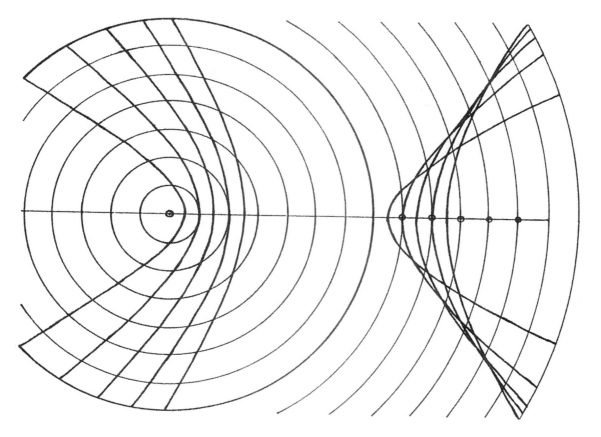

Fig. 120

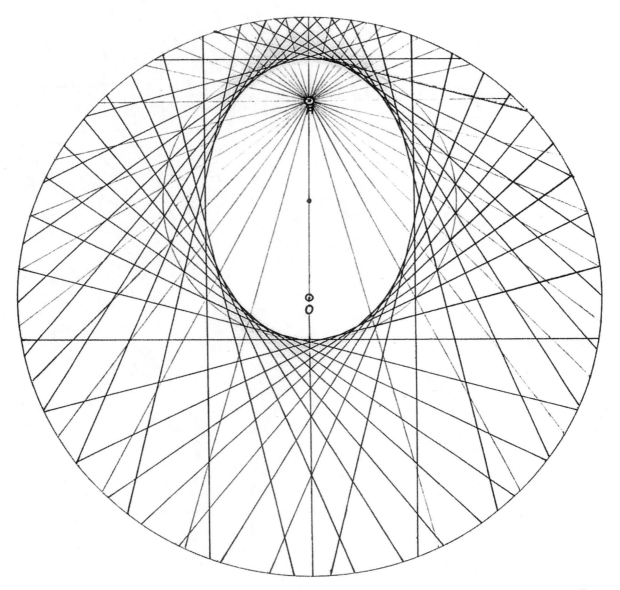

Fig. 121

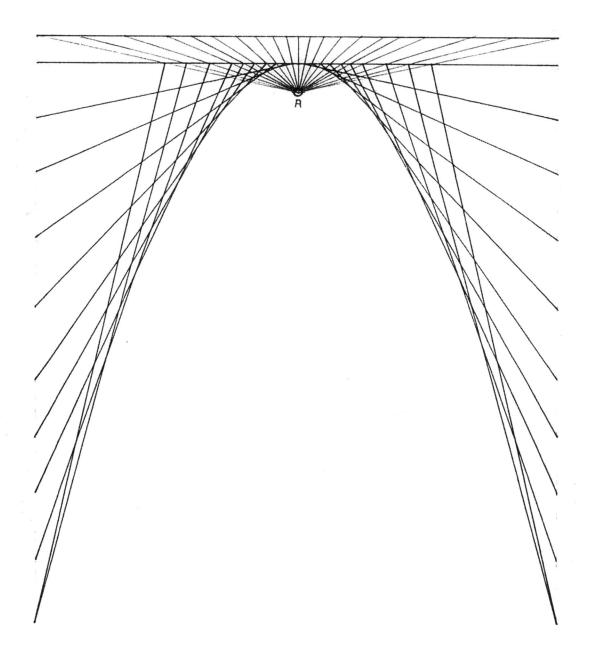

Fig. 122

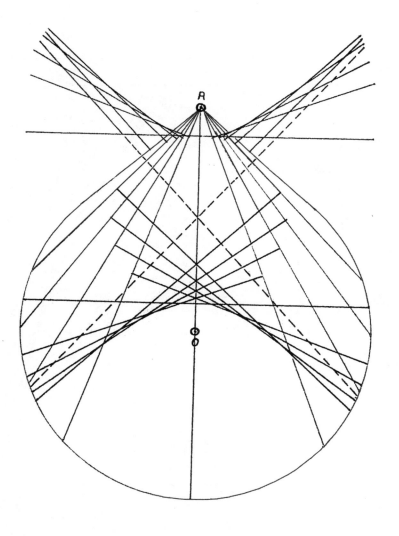

Fig. 123

In several of these drawings, concentric circles are the basis of the construction. The children may be reminded that in their early lessons in geometry they drew such families of circles (see Chapter 3).

So far, all our constructions of the conic section curves have been pointwise. There now follow linewise constructions of an ellipse (Fig. 121), a parabola (Fig. 122), and a hyperbola (Fig. 123). In Fig. 121, a point A is taken inside a circle (center O), and rays are drawn from A to the circumference. The midpoints of these rays are found (these midpoints lie on a circle whose center is the midpoint of AO), and from these points lines are drawn at right angles to the rays. These lines mold the form of the ellipse as shown. The construction is essentially the same for the parabola and hyperbola. In the case of the parabola, the circle is replaced by a straight line, the rays are drawn from point A to this line, and the right bisectors of these rays are drawn. For the hyperbola, the point A is taken outside the circle, rays are drawn to the circumference, their shorter and longer distances are bisected, and the right bisectors are drawn as before. It is interesting to notice that the right bisectors of the two tangent rays are the asymptotes of the hyperbola.

In the next chapter on projective geometry, further examples of linewise constructions of conics will be given.

Having become quite familiar with the constructions and forms of these three curves, we are now in a position to consider some of their fundamental geometrical properties. We shall do this, first, in relation to the significance of their names, and again we shall see in what follows that there is a certain unity preserved. From this point of view, the parabola will be considered first.

The Parabola

Where do we find this curve expressed in nature? As a static form it is rarely to be found. However, as a form expressed in movement, we recognize it at once in the flight of a ball or a stone thrown into the air or again in the path of a jet of water or the majestic curve of a waterfall. In mechanical terms, the parabola arises if a heavy body (a ball, stone, or water), which in the absence of any other force would move in a straight line with constant velocity, is also acted upon by the force of gravity. (It should, of course, be noted that the parabolic curve obtained by such a natural phenomenon is only approximate because there is always the third force of the air resistance acting.) Now the usual geometrical definition of a parabola refers to the curve as the path of a moving point as follows: A *parabola is the locus of a point that moves so that the ratio of its distances from a fixed point (F) and a fixed straight line (L) is always equal to unity.* In other words, every point on the curve is the same distance from the fixed point (F) as it is from the straight line (L), and we see that our construction of the parabola is exactly according to this definition. The fixed point F is called the focus and the fixed straight line L, the directrix, and the straight line passing through the focus and perpendicular to the directrix is called the axis of the curve. If P is any point on the curve (Fig. 124) and f and n its distances from the focus and the directrix respectively, then by the definition $f/n = 1$, or $f = n$. Now we have seen that a parabola is one of the conic section curves shown when a cone is cut by a plane parallel to one of its sides, that is, by a plane that makes an angle with the base of the cone equal to the base angle of the cone. Here again, this definition of the curve speaks of an equality, this time in the sphere of angular measurement. The name *parabola* comes from the Greek work *paraballo,* which means to throw or set beside, to compare. It is, of course, the same word as *parable,* which is a comparison, a similitude, or, simply expressed, an "earthly story with a heavenly meaning." The word in its various forms implies an equality.

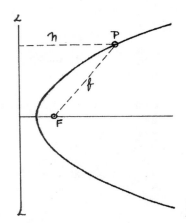

Fig. 124

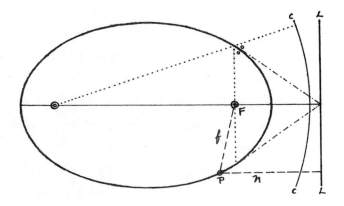

Fig. 125

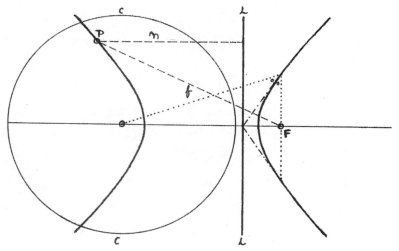

Fig. 126

The Ellipse

The corresponding locus definition of an ellipse is as follows: *An ellipse is the locus of a point that moves so that the ratio of its distances from a fixed point (F) and a fixed straight line (L) is a constant less than unity.* The ellipse in Fig. 125 is constructed as before with reference to a fixed point F and a circle C. The fixed straight line (L)—the directrix—of the above definition may be determined by drawing tangents to the ellipse at the extremities of a focal chord. These tangents intersect on the directrix. If P is any point on the curve and f and n its distances from the focus and the directrix, respectively, then by the definition, f/n = constant < n. (For the ellipse in Fig. 125, this constant is approximately 0.78.) To obtain an ellipse by cutting a cone, the cutting plane must make an angle with the base of the cone less than the base angle of the cone. The name *ellipse* comes from the Greek word *ellipsis,* meaning a "leaving behind," an omission, a deficiency. In grammar, an ellipsis is a figure of speech in which one or more words are left out, although their meaning is implied by the rest of the sentence.

The Hyperbola

For this curve, the corresponding definition states that: *a hyperbola is the locus of a point that moves so that the ratio of its distances from a fixed point (F) and a fixed straight line (L) is a constant greater than unity.* The hyperbola in Fig. 126 was constructed with reference to a point and a circle, and the straight line (L)—the directrix—was determined as for the ellipse. Again, if P is any point on the curve and f and n its distances from the focus and the directrix, respectively, then by the definition: f/n = constant > n (for the hyperbola in Fig. 126, this constant is approximately 1.4.) The cutting plane of a cone to give a hyperbola must make an angle with the base of the cone greater than the base angle of the cone. The word *hyperbola* comes from the Greek *hyperbole,* which means a "throwing beyond," an "over-shooting," an excess. In rhetoric, hyperbole is a figure of speech in which exaggerated terms are used as a means of emphasis, for example, when we say "a thousand apologies!"

To sum up then, we may say that the ellipse is a curve of deficiency, the parabola is a curve of equality, and the hyperbola is a curve of exaggeration, relating these words to the ratios of two lengths, these ratios being less than, equal to, and greater than unity, respectively.

In a previous chapter on the four rules of arithmetic treated geometrically, we "discovered" two of these conic section curves, the ellipse and the hyperbola, as "pictures" of the processes of addition and subtraction, respectively. We shall now proceed to show that there is a common law for all three curves in that they may be considered as curves of addition, and that they are three metamorphoses of one geometrical form. In order to understand this fully, we must now digress somewhat and consider certain fundamental concepts that belong to modern geometry.

We will begin with a consideration of parallelism. According to the Euclid definition, two straight lines are parallel if, when drawn in the same plane, they do not meet, however far they are produced in either direction. Now side-by-side with this, we have the axiomatic statement that any two straight lines drawn in a plane cut one another in one point and one point only. The definition of parallelism, though still perhaps the best for use in elementary geometry, is clearly at variance with this axiom, and such a contradiction is inconsistent with the logic and law of mathematical thinking. It is a definition applicable to a limited conception of space. A new definition of parallel lines, suggested by the principle of continuity and essential for an understanding of more advanced geometrical conceptions, was first given by Kepler (1604) and Desargues (1639): Lines, drawn in the same plane, are parallel if they have the same infinitely distant point in common. In other words, parallel lines meet in infinity. We see that, according to this definition, parallel lines are no exception to the statement that two straight lines drawn in a plane cut one another in one point and one point only. Furthermore, we shall see in what follows, and especially in the next chapter on projective geometry, that this conception of parallelism explains many phenomena of the geometrical nature of space which otherwise are quite baffling, and it also brings a wonderful consistency and order into geometry. We therefore have to consider geometry in relation to infinity.

The term *infinity* in mathematics, expressed simply, means "greater than any conceivable quantity" and is denoted by the symbol "∞." The *Encyclopaedia Britannica* (14th edition) says of infinity: "a term used in mathematics, philosophy, and theology with various meanings which are apt to cross each other and cause confusion." This confusion arises because our ordinary thinking based on sense perception is limited to the finite world. To grasp the infinite, a quality of imagination must penetrate our thinking—"a mathematical imagination" that will carry our thinking beyond the limits of sense-perceptible space.

Now the statement that "parallel lines meet in infinity" implies another imagination; that is, that there is one point at infinity on a straight line. Thus if we imagine a horizontal straight line continued infinitely far to the right and again infinitely far to the left, we do not come to two infinitely distant points, but only one. Let us now see what evidence there is for such a paradoxical statement; real imagination often presents the human intellect with paradoxes. Consider two straight lines drawn in a plane, one of them fixed and the other turning on an axis through O perpendicular to the plane (Fig. 127). Suppose the rotating line first cuts the fixed line at right angles in point P, and then we begin to turn it in an anti-clockwise direction and watch the passage of the cutting point P as it moves along the fixed line. As the rotation continues, the point P moves further and further away off the paper to the right until the moving line has turned through a right angle. Then as the turning continues beyond a right angle, the point P appears far away along the fixed line to the left until we see it coming back onto the paper and reaching its starting point when the rotation of the line has passed through two right angles. We can thus follow the continuous movement of P in our imagination, and we notice that it always moves in the same direction, provided that the moving line rotates in the same direction. We may now ask the question: Where is the point P at the moment when the moving line has rotated through exactly one right angle, that is, when it is parallel to the fixed line? It is then infinitely far away, and this point infinitely far to the right is one and the same as the point infinitely far to the left.

Another demonstration of the reality of this fundamental idea is to be found in the sphere of optics. If we follow the movement of the image of an object viewed in a concave spherical mirror, we can come to no other conclusion than that there is one point at infinity on a straight line. The series of drawings in Fig. 128 illustrates, by the well-known construction, the position of the image of an object as the latter moves towards the mirror. C is the center of curvature of the mirror and F its optical focus, which is half-way between the mirror and C. (See any elementary textbook on optics.) If I stand in front of a large concave spherical mirror (e.g., an old searchlight reflector) and walk slowly towards the mirror holding my hand well in front of my head and watching the movement of the image of my hand, this is what I observe: the image-hand (I), which is smaller than the object-hand (O) and is inverted, moves out from in front of the mirror to meet me (Fig. 128a, b). At one position on my "journey," I touch the image hand, which has "grown" to the same size as the object hand—I have the uncanny experience of shaking hands with my own image! This occurs when my hand is at the center of curvature of the mirror (Fig. 128c). As I now walk very slowly forward, I see the image hand still coming towards me and growing bigger. I lose sight of it, of course, as it moves away behind me, although another observer standing behind me can see it passing over my shoulder (Fig. 128d, e). When I now move forward between the focus and the mirror, I see the image hand coming to meet me from behind the mirror, erect and magnified but diminishing to the same size as the object hand when the latter reaches the mirror (Fig. 128g, h). When the object hand is at the focus (F), the image is at infinity (note the parallel rays). Notice that the image moves always in the same direction, and at one moment, when the object is just approaching the focus, it is a long way in front of the mirror, and the next moment, when the object has just passed the focus, it appears a long way behind the mirror. How can the image have "traveled" from far away in front of the mirror to far away behind it? Can it have turned round and come back? No, because then the observer would have seen it during its passage between his eye and the mirror, moving in the opposite direction. As he does not do so, there is only one conclusion we can make from our "experiment," that is, that the image makes a continuous journey along the straight-line axis of the mirror, through the point at infinity, and back from the same point at infinity behind the mirror. We see how this is entirely in accordance with the paradoxical statement of Desargues that a straight line is a circle of infinite radius and that therefore the two extremities of a straight line meet at infinity. Two interesting observations may be noted: The image changes from being erect to being inverted in passing through the point at infinity, and during its journey, it never appears between the mirror and the focus. This is "forbidden ground"; in the case of a convex mirror, however, the only "territory" the image can occupy is just between the mirror and its focus, which in this case is behind the mirror. Lewis Carroll knew what he was doing when he wrote *Alice's Adventures in Wonderland* and *Through the Looking Glass;* the Rev. C. L. Dodgson was a mathematical don at Christ Church, Oxford, and his stories contain imaginations of geometry in relation to infinity.

The reader may well think that these ideas are too difficult for even older children to assimilate. From long experience the author finds that this is not so. In fact, they find less difficulty than grownups because their imagination is generally more alive. Also the mirror experiment will not merely be described to them, but each member of the class will experience the phenomenon for him or herself. Again, very interesting and helpful discussions can arise with the children. It will become evident that to our ordinary thinking, bound as it is to our physical sense impressions, parallel lines simply do not meet, and one has to go beyond ordinary thinking into the realm of imagination to "perceive" that they do meet in infinity. The author has sometimes suggested the following ideas to his class: The boundary between the finite physical world and the realm of the infinite is a boundary of human consciousness. If one were able to awake from sleep with a clear consciousness of where, as an individuality, one had been during sleep, then one would have a real waking knowledge of the world of infinity and of the laws of that world. With these ideas of modern geometry in mind, we may now continue our study of the conic section curves.

The ellipse in Fig. 129 was drawn by the well-known method of a loop of cotton around two pins stuck into the paper at F_1 and F_2. This method of construction leads directly to the realization that the ellipse is a curve of addition. Thus, if P and P' are any two points on the ellipse and r_1 and r_2 and r_1' and r_2' are their distances from the foci F_1 and F_2 respectively, then

$$r_1 + r_2 = r_1' + r_2' = \text{constant}$$

We also see this additive property of the ellipse quite clearly from the method of construction used originally (see Fig. 110), where the curve appeared as a line of symmetry between a point (F_1) and a circle (C) enclosing the point. Thus any point P_1 on the curve is equidistant from F_1, and the circle C, that is, f = c for all positions of P_1. Therefore $OP_1 + f = OP_1 + c$ = radius of circle C = constant.

Now keeping focus F_1 fixed, we move the focus F_2 out along the axis of the curve and obtain larger and larger ellipses. When F_2 has been moved infinitely far from F_1, then the ellipse has become metamorphosed into a parabola (Fig. 130). We may say that a parabola is an ellipse with one focus at infinity. Thus unlike the ellipse, the parabola appears at first sight to be an open curve. Again we choose any two points, P and P', on the curve and join them to the foci, F_1 and F_2. The two lines parallel to the axis are the focal rays joining the points to F_2 at infinity. The length F_1P is stepped off along F_1P' by drawing arc PL, and the length F_2P' is stepped off along F_2P by drawing the perpendicular P'M, which is also the arc of a circle with center F_2 at infinity. Now the two heavily drawn lines PM and P'L are found to be equal, which means that the focal ray r_1' is longer than r_1 by the same amount that r_2' is shorter than r_2.

Thus for the parabola, $r_1 + r_2 = PF_1 + PM + M$ to ∞

And $\qquad\qquad\qquad r_1' + r_2' = LF_1 + P'L + P'$ to ∞

But $\qquad\qquad\qquad PF_1 = LF_1, PM = P'L + M$ to $\infty = P'$ to ∞

$\qquad\qquad\qquad\qquad r_1 + r_2 = r_1' + r_2' =$ constant

So the law of the parabola in relation to infinity is the same as that of the ellipse; that is, it is also a curve of addition. Now if we continue the journey of the focus F_2 back from infinity (as it were, from the other side) into finite space, we obtain the third metamorphosis, as hyperbola (Fig. 131). We may perhaps say that while the focus F_2 has been making this continuous infinite journey through the point at infinity on the axis and back again (though not "back again" by turning round), the ellipse has changed into a parabola and then has gone through a kind of turning inside-out process together with a twist to appear once again on the paper in the form of a hyperbola. This "turning inside out" and "twisting" may be imagined more easily if we follow round each curve, say in a clockwise direction as shown by the arrows in the three drawings, and at the same time consider the space inside and outside the curves.

In the case of the ellipse, there is no difficulty in doing this as it is a closed curve containing a finite space, and therefore its inside and outside are quite obvious and, of course, the two foci are inside the curve. The mathematics of the ellipse is also easy to follow because the curve is "all there," and all the distances concerned with it are finite. The parabola presents more difficulty because it is closed only if we include infinity in our conception; but even here, we can follow round the curve in our imagination and experience the inside and the outside quite easily. To try and follow the curving of the hyperbola, we use two "guiding lines," which form a cross as shown in Fig. 131. These two lines are called asymptotes (from the Greek word *asymptotos,* meaning "not falling together"), and their position relative to the hyperbola is found, as shown, by drawing the circle with diameter F_1F_2 and the two tangents at the ends of the curve. The asymptotes are a limiting case of the conic sections, obtained by a plane cutting a cone along its axis; they are the slant edges of the cone when the cone is viewed in elevation. Now let us imagine a point moving along the right wing of the hyperbola in the northeast direction (see Fig. 131). The curve approaches the asymptote ever more closely the further it goes but touches it only at infinity (hence the meaning of the word *asymptote*). It then returns on to the paper from the southwest, seemingly on the other side of the same asymptote, curves round towards the northwest, approaching the second asymptote more and more nearly, touches it at infinity and returns from the southeast. again on the other side, and curves once more towards the northeast to complete its continuous course. The space where the asymptotes lie is outside the hyperbola, and its inside, embracing finite and infinite space, is where the foci lie. Hence, when we join the two points P and P_1 to focus F_1 we must do so through the inside space as we did in the case of the ellipse and the parabola. Thus P and P_1 are joined to F_1 via infinity. Then by a construction similar to that used with the parabola, we see that the focal ray

r_1' is longer than r_1 (both via infinity) by the same amount (the equal heavily drawn lines PM and P'L) that r_2' is shorter than r_2.

Thus for the hyperbola,
$$r_1 + r_2 = PF_1 \text{ (via } \infty) + PM + MF_2$$
$$r_1' + r_2' = LF' \text{ (via } \infty) + P'L + P'F_2$$

But $\quad PF_1 \text{ (via } \infty) = LF_1 \text{ (via } \infty), PM = P'L, \text{ and } MF_2 = P'F_2$

Therefore, $\quad r_1 + r_2 = r_1' + r_2' = \text{constant}$

Thus the law of the hyperbola in relation to infinity is the same as that of the ellipse and the parabola. That is, it is also a curve of addition.

We can now understand why in Chapter 5, the hyperbola appeared as a picture of subtraction. We then considered one of the focal distances across the outside of the curve (focal distances *a* or *c* in the three drawings of Fig. 86, and this is of course a necessary substitute if we require to make an actual measurement of length. This substitute relationship of constant difference instead of constant sum may be shown as follows from our present drawing (Fig. 131) if we consider the two "dotted" lines joining P and P' to F_1 across the outside of the curve (n_1 and n_1, respectively):

$$n_1 - r_2 = PF_1 - (PM + MF_2) = PF_1 - PM - MF_2$$
$$n_1' - r_2' = (LF_1 - LP') - P'F_2 = LF_1 - LP' - P'F_2$$

But $\quad PF_1 = LF_1, PM = LP', \text{ and } MF_2 = P'F_2$

Therefore $\quad n_1 - r_2 = n_1' - r_2' = \text{constant}$

During their drawing of the conic section curves, the children will often ask the question as to the difference between a parabola and one branch of a hyperbola, since they both open out into the infinite. The difference is in the whole way in which the two curves go out into infinity, and the following drawings (Figs. 132 and 133) illustrate this.

In Fig. 132 are shown four identical parabolas and in Fig. 133 two identical rectangular hyperbolas (i.e., hyperbolas having their asymptotes at right angles), constructed by the same method as Figs. 109 and 111. The distances between the foci in the two sets of drawings are the same so that the figures are comparable. We see how the parabola curves gently and slowly away into the infinite while the hyperbola shoots off into infinity very rapidly. The cross formed by the four parabolas is an artistic form, while that formed by the two hyperbolas has more the appearance of part of a machine.

We have seen both in this chapter and in Chapter 5 that there are many interesting ways in which the teacher can construct and characterize the conic section curves and lead the children to understand them from different points of view.

What we have elaborated here in some detail enables us to see the wonderful unity of inner law expressing itself in these three curves, which outwardly appear in such differing forms. They are, in fact, three metamorphoses of a fundamental geometrical form: a unity revealing itself as a trinity, which is perhaps one of the greatest realities to be found in mankind's comprehension of himself and of the universe around him.[14] It may be added that such a fundamental reality belongs to a more imaginative and spiritual understanding of phenomena than is usual in our ordinary life of thought, and it is this we are attempting here in studying the conic section curves in relation to the infinite.

Let us consider this trinity of the conic section curves in relation to certain natural phenomena. The ellipse and its limiting form, the circle, are to be found everywhere as fundamental forms in the creations of nature. From the spherical form of the earth to the tiny spherical or ellipsoidal shapes of an organic cell structure, we find this curve universally expressed. It is a basic form in the material world. Now the hyperbola, both in its geometrical form and in its expression in natural phenomena, is the polar opposite of the ellipse. Since it is an open curve embracing infinite space, it is obvious that it is not to be found as a form assumed by material substance. It is, however, to be found in a more subtle way in relation to the least dense state of matter, that is, a gas, and just in connection with the effect of the still more imponderable element of heat on the gas. Unlike the solid and the liquid, all gases expand under the influence of heat by the same amount.

Every true gas expands $\frac{1}{273}$ of its volume at 0° C for every 1° C rise in temperature. Each particular solid or liquid has its own individual coefficient of expansion. The hindrance to the expansion of a gas is given by the wall of the gas container, and the expression of this hindrance is the pressure that the gas exerts. In this pressure one can, so to say, read off the "expansion-striving" of the gas.

Now the physical law according to which, at constant temperature, the pressure of a gas varies relative to the space it occupies is the well-known Boyle's Law, which states that, at constant temperature, the product of the pressure and the volume of a given mass of gas remains constant. That is, PV = constant. The graphical picture of this law, plotted with rectangular Cartesian coordinates is one wing of a rectangular hyperbola. Yet again, we may refer to two earlier drawings where, although the curves have not actually been drawn, the rectangular hyperbola can readily be seen. Fig. 26 in Chapter 3 shows a family of rectangles of equal area (length × breadth = constant), which is the law of the hyperbola. The curve may be obtained by joining the corners of the rectangles by a smooth curve. In Fig. 27 we have a family of rhombuses of equal area, and here their sides envelop hyperbolic curves. This gives a linewise construction for a hyperbola, whereas the constant area rectangle method is pointwise. So we may say that the hyperbola appears as the geometrical expression of a law of nature concerned with the effect of a nonmaterial entity (heat) on the least dense state of matter (a gas). How polarically different this is from the appearance of the ellipse in the material, substantial forms that Nature creates! Thus the ellipse and the hyperbola are to one another as two members of a polarity, both as regards their geometrical properties and their manifestation in natural phenomena, though they are a unity through the fact that they express a common law in that they are both curves of addition. In between them as a balancing, compensating element stands the parabola. The extremes of ellipsis and hyperbola find their balance in the parabola. The ellipse belongs entirely to finite space and is fully comprehended within that space; the hyperbola has its "being" in infinite space and is only understood in relation to infinity: it merely shoots its "extremities" into finite space while its real "body" remains hidden from physical perception. The parabola in a gentle, harmonious way binds the finite and the infinite together. We comprehend it partly in physical and partly in infinite space, and we can follow its curving line in our imagination without difficulty, whereas to do so in the case of the hyperbola requires what we may call a kind of "mental gymnastics" in the sphere of the imagination. Now in the world of physical phenomena we find that the parabola manifests itself in just this mediating role between the finite and the infinite—between the earth and the heavens. Again, since it is not a closed curve, we do not find it as a form enclosing substantial matter as in the case of the ellipse, or rather ellipsoid, but as a "container" for super-earthly forces and energies. A parabolic mirror, directed with its axis towards the sun, reflects the light and the heat of the sun exactly to its focal point. The light and heat coming from infinite space are concentrated, condensed at this one point in finite space. (The German word for focus *Brennpunkt*, burning point, is very descriptive of this phenomenon.) If, on the other hand, we consider a parabolic curve with its axis pointing the opposite way toward the center of the earth, then we have a picture of the flight of a solid body through the air, or the course of a jet of water spurting out of a hose pipe. Thus the parabola conceals within itself cosmic and earthly laws; it belongs both to the sunlight and to earthly gravity. The following drawings (Figs. 134 and 135) illustrate these two aspects of the parabola—the reflection of the sun's light to a focus point and the path of a body projected from the earth's surface at an angle of 45° with an initial velocity of 128 feet per second. (As pointed out earlier, the projectile would actually fall short of the distance shown owing to the resistance of the air.) It may be observed that light raying out from one focus of an ellipse would be reflected to and concentrated exactly at the other focus, while from the focus of a parabola, light after reflection streams out in parallel rays into infinite space. The reflector of the modern motor-car lamp is parabolic, as also are the mirrors in reflecting microscopes and telescopes with which we obtain the infinite within our finite field of sight. There is an interesting phenomenon that may be observed when light is reflected from a circular or spherical surface; we may become aware of this if we look on to the surface of a cup of tea when the sun is shining on the white inside surface of the cup. We then see a beautiful curve of light against the dark background of the tea.

This phenomenon is illustrated in Fig. 136, the curve being shown in thick line. This is known as a caustic curve, with its cusp halfway between the mirror and its center of curvature. It is formed by the reflected light that is not all concentrated at the focus as in the case of a parabolic mirror. Only the light falling on a small aperture of mirror near the pole is reflected to the focus, the light outside this being focussed progressively nearer to the mirror. This explains why, when a spherical mirror is used instead of a parabolic one, the light must only be allowed to fall on a small aperture of the mirror if we require a sharp focussing of the light. Otherwise we get the phenomenon shown here which is called "spherical aberration." The parabolic mirror with the same focus is shown by the broken line. It will be noticed that the two mirrors very nearly coincide over a small aperture, and it is therefore for this small aperture that the spherical mirror gives a sharp focus since here it obeys the law of the parabolic mirror.

Other relationships of these curves to natural phenomena and especially to the being of man are of the greatest interest and importance. They would however, take us into too much detail in one particular sphere of our studies, and the reader is therefore referred to an article by Ernst Bindel, "Die Kegelschnitte in menschengemasser Behandlung," published in the *Jahrbuch der naturwissen-schaftlichen Sektion der Freien Hochschule für Geisteswissenschaft am Goetheanum,* Dornach, near Basel, Switzerland.

One final example of the occurrence of the conic section curves in nature may, however, be given. It concerns the movements of the Sun and the stars in relation to the earth. By day and by night, countless shadows of trees, buildings, and other objects are cast over the surface of the earth by the light of Sun, moon, and stars, and the ends of these shadows, as they move, trace one or other of the conic section curves. At the poles where all the heavenly bodies describe horizontal circles, the "shadow curve" traced out on the earth is itself a circle. Travelling away from the poles we find the shadow curve changed to an ellipse. In these polar regions, there are quite a few stars that rise and set, and their shadow curve appears as a hyperbola, while those stars that exactly touch the horizon in the course of their daily journal form a parabola as their shadow curve. Thus in the polar regions of the earth where the stars are mostly circling above the horizon the whole of each day, there is a great preponderance of elliptical shadow curves traced out on the surface of the earth. (The word *shadow* here does not always signify a visible shadow; only the Sun and the moon cast visible shadows.) At the equator, on the other hand, where the stars daily rise and set, all the shadow curves are hyperbolic. Here again, the fundamental polarity of ellipse and hyperbola is expressed.

The indications that have been given here will serve to show that the conic section curves "belong" in a profound manner to the very nature of space and therefore to spatial phenomena. Moreover these spatial phenomena are really only to be understood when we relate them to an infinite space. This may be expressed in more general terms by saying that behind all physical phenomena there is a spiritual reality. There is a constant interplay between earthly and heavenly forces. The hyperbola is no less a reality than is the ellipse. "God is eternally geometrizing."

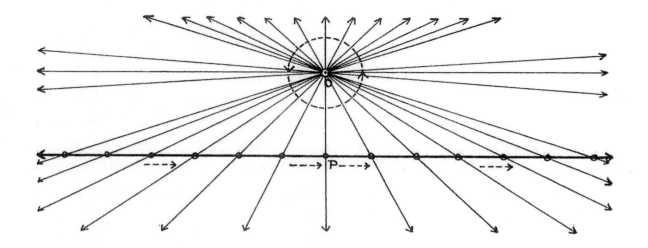

Fig. 127

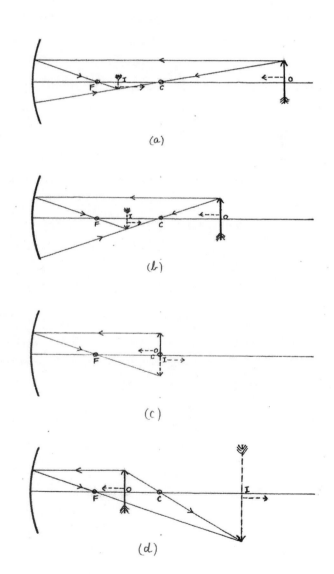

Fig. 128 a, b, c, d

174

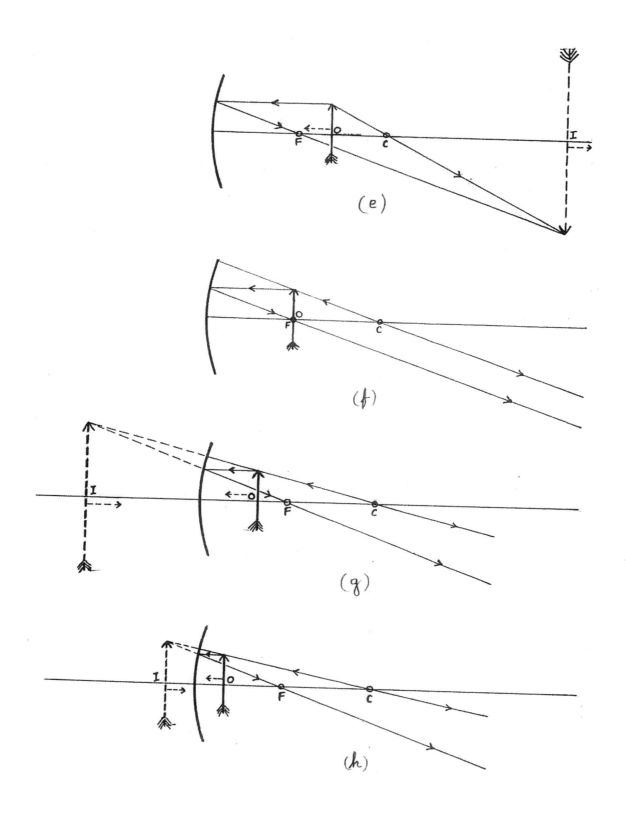

Fig. 128 e, f, g, h

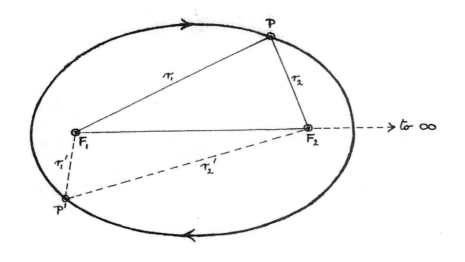

Fig. 129

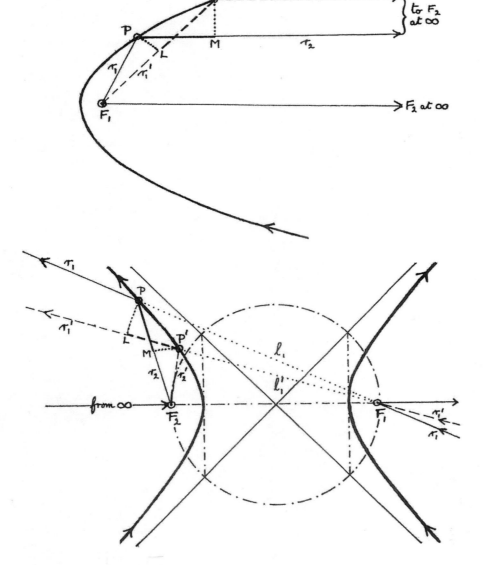

Fig. 130

Fig. 131

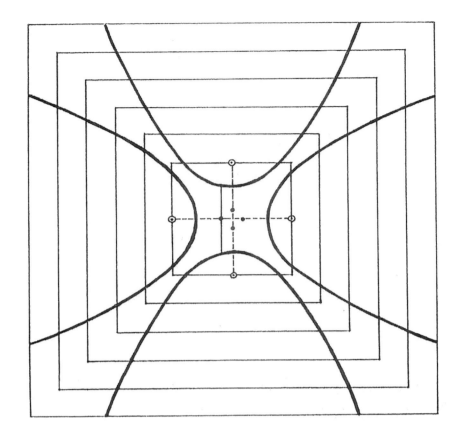

Fig. 132

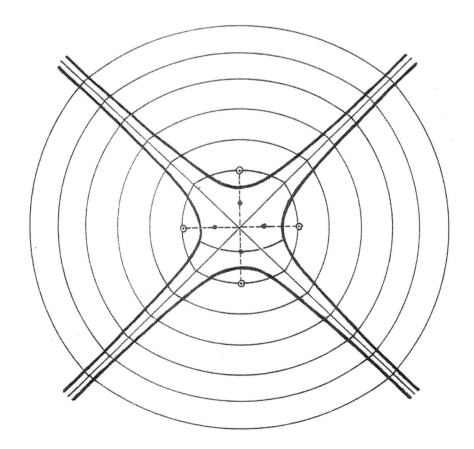

Fig. 133

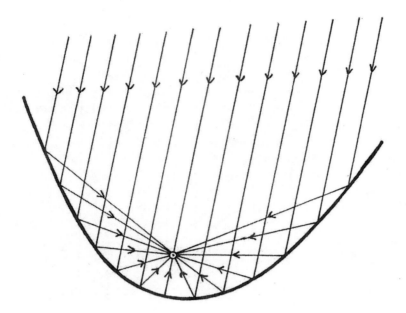

Fig. 134

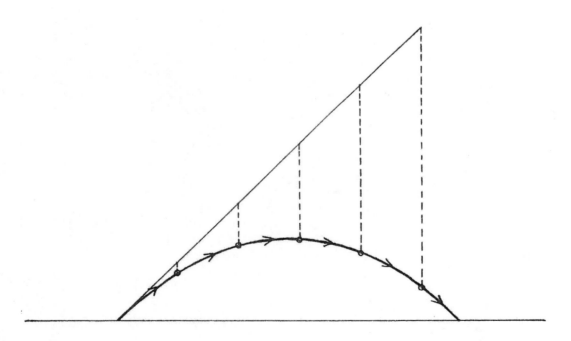

Fig. 135

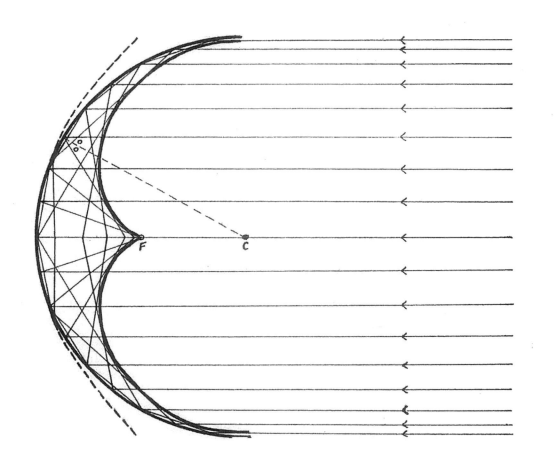

Fig. 136

Chapter 8

Projective Geometry

Wherever we look we see a perspective view. An expert artist or craftsman standing before a group of objects records exactly what he sees; the result is a perspective drawing. The well-known painting in the National Gallery, London, of the avenue of trees by the Dutch artist Hobbema (1638–1709) immediately comes to mind. It was only at the beginning of the modern age that artists and scientists awakened to an experience of spatial perspective. Filippo Brunelleschi (1377–1446), the leader of a group of young Florentine artists, was not only the initiator of Renaissance architecture but made the momentous discovery of perspective in the field of art. Other famous artists such as Leonardo da Vinci (1452–1519) and Albrecht Dürer (1471–1529) developed it to a very high degree of perfection. One of the first paintings made according to the mathematical rules of perspective is a wall painting in the church of St. Maria Novella, Florence, and represents the Holy Trinity with the Virgin and St. John at the foot of the Cross and the donors, a merchant and his wife, kneeling outside (Plate 5). The artist was Tommaso Guidi (1401–1428), nicknamed Masaccio, meaning "clumsy Thomas." As E. H. Gombrich says in *The Story of Art*: "He must have been an extraordinary genius, for we know that he died when hardly twenty-eight years of age, and that, by that time, he had already brought about a complete revolution in painting.... We can imagine how amazed the Florentines must have been when this wall painting was unveiled and seemed to have made a hole in the wall through which they could look into a new chapel in Brunelleschi's modern style."

This was indeed the beginning of a new age. The individuality of man takes a great step forward; he made quite a new approach to the external world, and he looked out into the wide surrounding spaces and saw objects and phenomena in a different way. The artist was the forerunner of this new way of looking at the world—perhaps the artist is always the forerunner of a new age in man's development. Then came the fifteenth and sixteenth centuries, the age of the great discoveries and explorations of the earth: Henry the Navigator, Christopher Columbus, Vasco da Gama, Ferdinand Magellan, and others.

While these men were exploring the world, others, no less adventurous and courageous, were beginning to explore the material substance and phenomena of the earth and of the heavens in quite a new way: Copernicus, Galileo, Francis Bacon, and later in the seventeenth century, Sir Isaac Newton, were bringing in the age of natural science, which has increasingly dominated the lives of people the world over down to our own day and brought to us such wonderful achievements.

This new and independent man standing with his feet firmly planted on the earth and looking out upon the world with a penetrating gaze is wonderfully portrayed in Donatello's marble statue of St. George from the Church of Orsanmichele, Florence (Plate 6). Donatello (1386–1466) was the greatest sculptor of Brunelleschi's circle. Referring to this statue Gombrich says, "If we think back to the Gothic statues outside the great cathedrals, we realize how completely Donatello broke with the past. These Gothic statues hovered at the sides of the porches in calm and solemn rows looking like beings from a different world. Donatello's St. George stands firmly on the ground, his feet planted resolutely on the earth as if he were determined not to yield an inch. His face has none of the vague and serene beauty of medieval saints; it is all energy and concentration. He seems to watch the approach of the monster and to take its measure, his hands resting on his shield, his whole attitude tense with defiant determination."[15]

It is indeed to the great Italian masters of painting that we owe this "conquest" of the delineation of space in their works of art. Leonardo da Vinci (1452–1519) (see Plate 3) and Raphael (1483–1520), to name only two, painted in the most accurate perspective. Among the northern artists, Dürer (1471–1528) was also a master of perspective drawing and wrote a textbook on the subject. Although sometimes his perspective drawing was inaccurate, a fine example of his work is *St. Jerome in His Study* (Plate 7).

Plate 5

Plate 6

Plate 7

This practical art of perspective originating in the fifteenth century developed gradually into pure mathematical conceptions of the nature of space itself. Thus there arose what is now known as projective geometry, which, although it embraces the theory of perspective, goes far beyond this and illuminates for us the very way in which space is built up.

Early in the seventeenth century, the first real advance since the time of the ancient Greeks was made in geometry. There were two main streams: (1) the analytic stream associated with the name of Descartes, who published the first treatise on the subject in 1637, and (2) the synthetic stream, with the new principle of perspective. The early investigators in synthetic or projective geometry were Desargues (1593–1662), an architect and engineer of Lyons, and Pascal (1623–1662), the famous French philosopher and mathematician. Contemporary with these investigations was the constant research into the difficult problem of the classical Euclidean geometry, namely the axiom of parallelism, to which we have already referred in the previous chapter. The study of projective geometry reached mature development in the nineteenth century in the works of Jacob Steiner and others, and it became the basis of all geometry. In fact, the English mathematician Arthur Cayley said: "Projective geometry is *all* Geometry." In our day, there are a number of important works on the subject by leading mathematicians, and these are mentioned in the booklist at the end of this book. In this connection, special mention should be made of the valuable research work being carried out by George Adams[16] of the Goethean Science Foundation, Clent, Stourbridge, Worcestershire, and by L. Locher-Ernst of the Goetheanum, Dornach, Switzerland. Following indications given by Rudolf Steiner, Adams is showing "the high significance of the new geometry for a more spiritual idea of space, and for a world conception free from the bonds of näive materialistic fancy." He has already written several important books (see Selected List of Books), of which the last shows how the conceptions inherent in projective geometry give us a real understanding of plant forms and growth.

Both for its own sake and because it is "*all* Geometry" and therefore belongs to a new and more spiritual understanding of spatial form and phenomena, it should be an essential part of a school curriculum. Being a very wide subject, the teacher may have some difficulty in deciding where to begin and with what aspects he or she should deal and how to present them. What now follows has arisen out of the experience of the author in teaching this subject to boys and girls of 17 and 18 over a number of years. A further important aspect of this work in education is that it calls forth in the pupil a kind of thinking that has a strong imaginative quality. The "thought forms" of projective geometry are themselves of such a nature.

We may take our start from a consideration of point, line, and plane, the three ideal geometrical entities of which all forms are a synthesis. This also applies to curved forms. The older classical geometry and especially the Cartesian analytical geometry, concerning themselves essentially with fixed and rigid forms, regard the point as the fundamental entity, and the straight line and the plane are, generally speaking, considered as composite entities made up of innumerable points. This emphasis on the point is to be found again in the conceptions of modern physics with its centrifugal and centripetal forces, that is, forces acting away from and towards point centers. Cartesian geometry is thus a most efficient handmaid for the ideas of physical science. Projective geometry however does not consider the point as having such unique importance; point and plane are equally valid and of equal importance as fundamental entities of space. Thus we may still consider the plane as formed by innumerable points, but then we must regard the point also as composite and formed by an infinitude of planes that contain it. Or, vice versa, if we consider the point as a single whole, we may also regard the plane in the same way, that is, as a single entity. Between these two polar entities of point and plane lies the third, the straight line, which as it were mediates between them. Lines ray through space weaving the countless forms that occupy space. Now the line[17] itself may be considered in three ways: (1) as an individual, self-contained entity, (2) as made up of an infinite number of points, and (3) as made up of an infinite number of planes. Thus, if we consider the points of a line as its parts, we must also consider the planes of a line (i.e., the planes that have the line as their common axis) as its parts, since we are not regarding the point as having more validity and importance than the plane. Qualitatively we may see the point and the plane as the expressions of the great polarity of contraction and expansion, while the line, contracted in one dimension and expanded in the other, mediates between these polar opposites. The author is aware that in the customary treatment of the subject, point, line, and plane are not dealt with in this way in introducing projective geometry. For example, A. N. Whitehead and Veblen and Young begin their writings with axioms or assumptions in which the point is treated as the fundamental entity. The present method of approach leads, in the author's opinion, to a more imaginative conception of space and is also one that can be more readily grasped by the boy or girl of school age.

There is a fundamental principle of projective geometry to which we shall often refer, known as the principle of duality.[18] Actually we have come across it already in the dual relationship of the five regular solids (see Chapter 6). In every theorem of projective geometry concerned with points, lines, and planes and figures arising from them, if we change the word *point* into *plane* or *plane* into *point*, we obtain an equally valid statement of truth.[19] Each theorem is then said to be the dual of the other. The following are simple examples that are quite self-evident or axiomatic:

Any two *points* in space have one *line*, and one line only, in common (the straight line joining them).	Any two *planes* in space have one *line*, and one line only, in common. (When the planes are parallel this line is at infinity.)
Any three *points* in space, not in the same straight line, have a single *plane* in common.	Any three *planes* in space, not in the same straight line, have a single *point* in common.

As plane is to point, so point is to plane. Consider a sphere; then by infinite expansion the sphere becomes the plane at infinity, and by infinite contraction it becomes a point (the center).[20] From what has already been said it will be evident that the concept of infinity belongs essentially to this geometry. This concept of infinity has been referred to in some detail in the previous chapter. Here we shall be dealing mostly with the two-dimensional aspect of projective geometry, namely with the relationships of lines and points in a plane, where the most expanded entity is the line. So the principle of duality will concern points and lines: As line is to point so point is to line.

| Any two *lines* (drawn in a plane) determine a *point*. (If the lines are parallel, the point is at infinity.) | Any two *points* determine a *line*. |

So much then for some of the basic ideas by way of introduction.

Let us begin with a simple drawing, which is attributed to Pascal (Fig. 137).

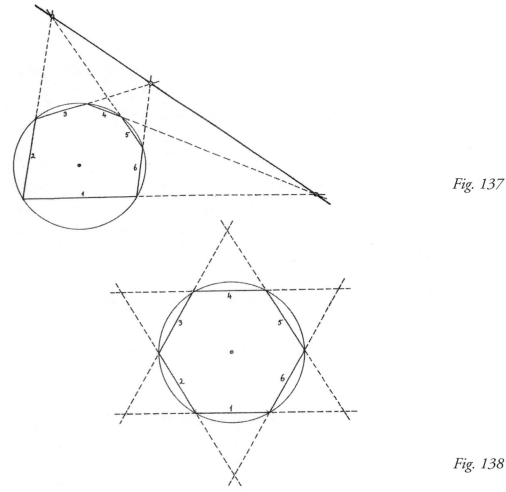

Fig. 137

Fig. 138

A hexagon is inscribed in a circle; the opposite pairs of sides; that is, sides 1,4; 2,5; and 3, 6, are produced to meet in three points. Then these three points always lie in a straight line (called Pascal's line). What a simple and remarkable relation of lines and points this is! There is no measurement of lines or angles, the hexagon is quite irregular, and yet the three points always yield a straight line. Truly it must belong to the very nature of space that this should be so! According to the particular hexagon drawn in the circle, Pascal's line may be "north, south, east, or west," and it may be nearer to the circle or further away— it may be anywhere. Now if the hexagon is regular (Fig. 138), the opposite pairs of sides meet in infinity (see Chapter 7). Therefore, the three points in which they meet are at infinity and thus Pascal's line is at infinity. Consider next a hexagon drawn outside a circle having its sides tangential to the circle, and change the above description of Fig. 137 so that the word *sides* becomes *points* and the word *points* becomes *lines*. Then we have a description of Fig. 139. A hexagon is circumscribed round a circle; the opposite pairs of points are joined, that is, points 1,4; 2,5; and 3,6, by three lines; then these three lines always lie in a point (called Brianchon's point after the French mathematician C. J. Brianchon, 1806). This is another example of the principle of duality. When the hexagon is regular, Brianchon's point is the center of the circle (Fig. 140). If we consider these two theorems of Pascal and Brianchon together, we see that the line at infinity (Pascal's line for a regular hexagon inside a circle) corresponds or is related to the point center (Brianchon's point for a regular hexagon outside a circle). This corresponds in plane geometry to what we have already referred to as the great polarity of point (infinite contraction) and plane (infinite expansion).

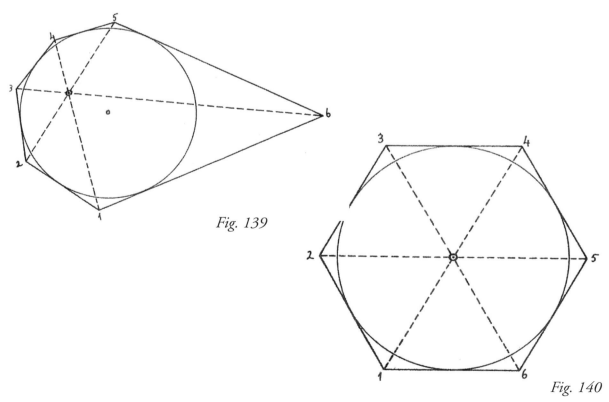

Fig. 139

Fig. 140

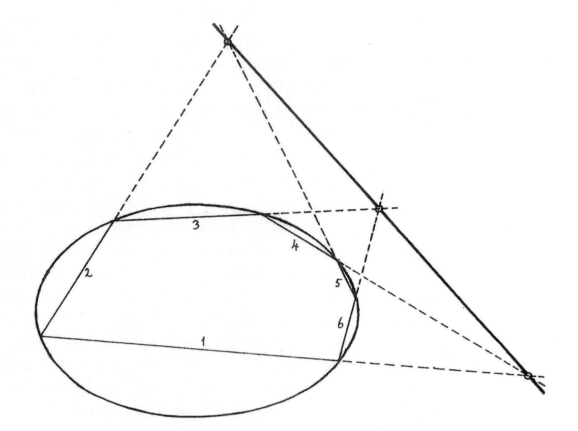

Fig. 141

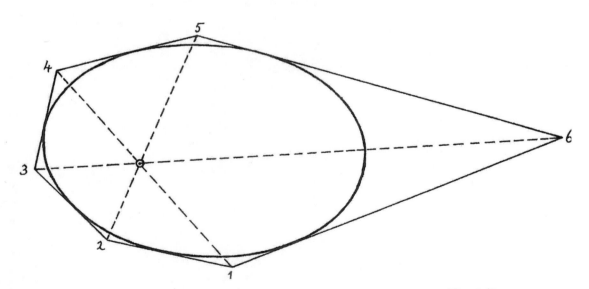

Fig. 142

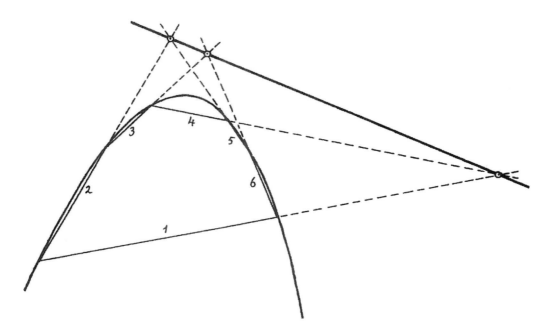

Fig. 143

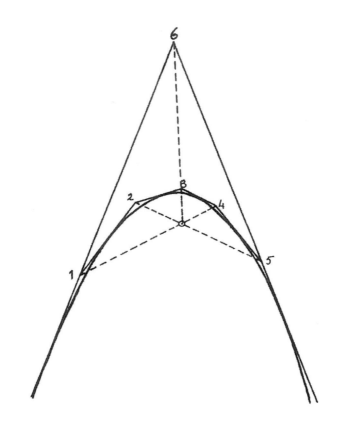

Fig. 144

191

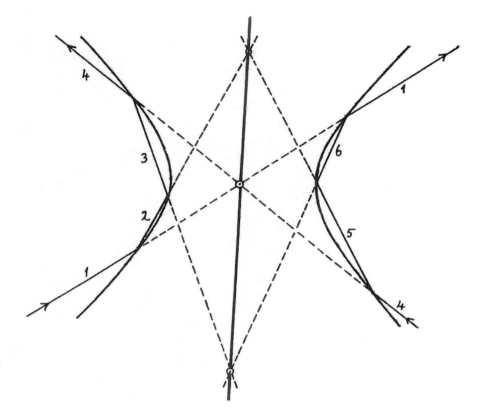

Fig. 145

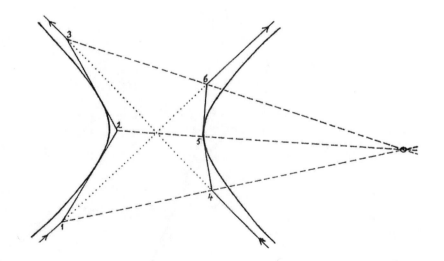

Fig. 146

Now we have seen in the last chapter that the circle is one of the conic section curves—it is what the ellipse becomes when the two foci coincide. We should therefore expect that since Pascal's and Brianchon's theorems express a pure relationship of lines and points in a plane, they would be valid for any of the conic sections. This is indeed the case, as the drawings for the ellipse, parabola, and hyperbola show. In the case of the ellipse, it is also evident if we look at Fig. 137 on the slant; the circle then appears as an ellipse, all the angles between the lines appear different, and yet the three points still remain in a straight line. This is, of course, a projection of Fig. 137. Similarly, Fig. 139 may also be projected to give Brianchon's point in an ellipse. Figs. 141 to 146 show Pascal's line and Brianchon's point for the ellipse, parabola, and hyperbola. In Fig. 144 where the hexagon is drawn outside the parabola, the angles 2, 3, and 4 are re-entrant (angles pointing inwards); it must be remembered that a hexagon is any figure made up of six lines. The hexagon drawn inside a hyperbola (Fig. 145) must be an open figure; sides 1 and 4 meet the curve from infinity. Fig. 146 shows the hexagon drawn outside a hyperbola; here two of the sides are parts of the asymptotes (the other parts being shown by dotted lines), which touch the curve in infinity. It should also be noticed how entirely consistent these drawings are; in each case Pascal's line is outside the curve and Brianchon's point is inside. Now it is also possible to draw the hexagons in such a way that Pascal's line is inside or cuts across the curve, and Brianchon's point is outside the curve. If we take six points at random on the circumference of a circle (Fig. 147a), one point can be joined to each of the others in five ways. This will give altogether:

$$5 \times 4 \times 3 \times 2 \times 1 = 120 \text{ hexagons}$$

but half of these will be the same, so the total number of different possibilities will be 60. Four of these are shown as examples (Fig. 147). Of course, 1 of the 60 will be when the points are joined in order of their arrangement round the circle, and then we have the case of Fig. 137. Pascal called these configurations of six lines drawn in a conic "the mystic hexagram." Fig. 148 is an example of one possible hexagon drawn outside a circle so as to give Brianchon's point outside.

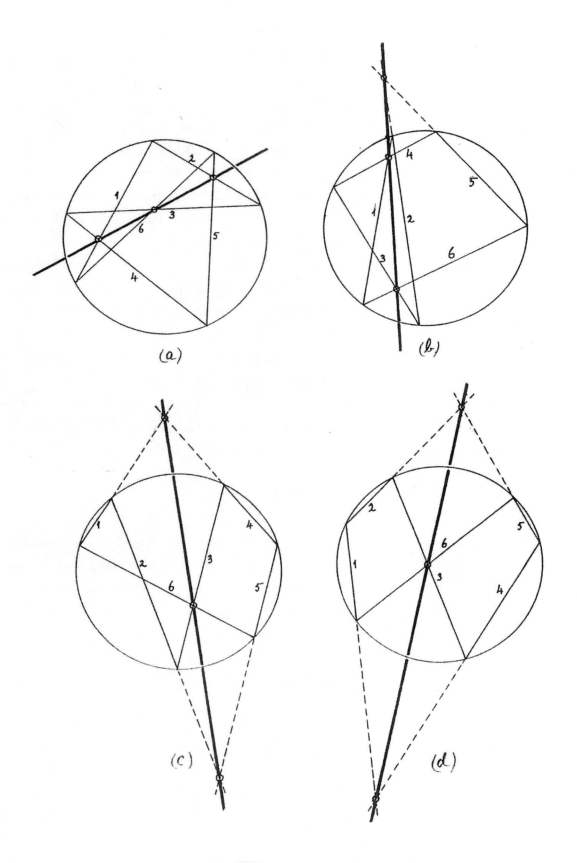

Fig. 147

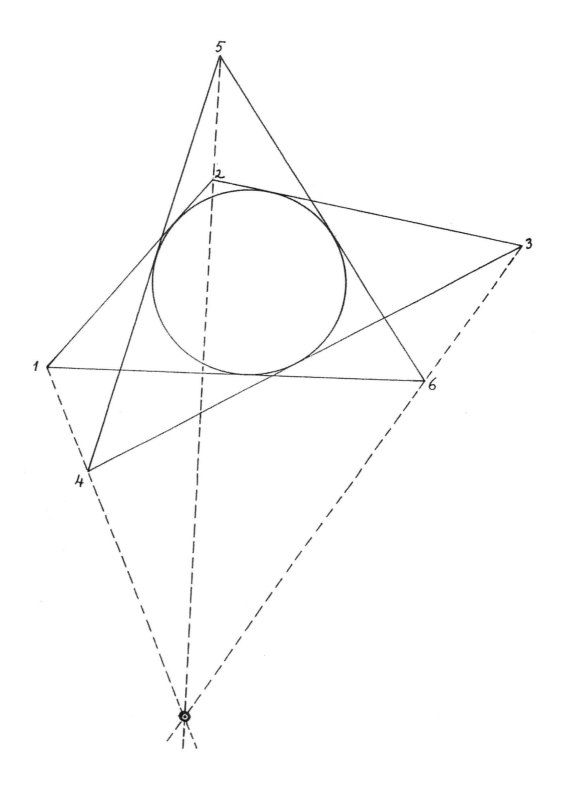

Fig. 148

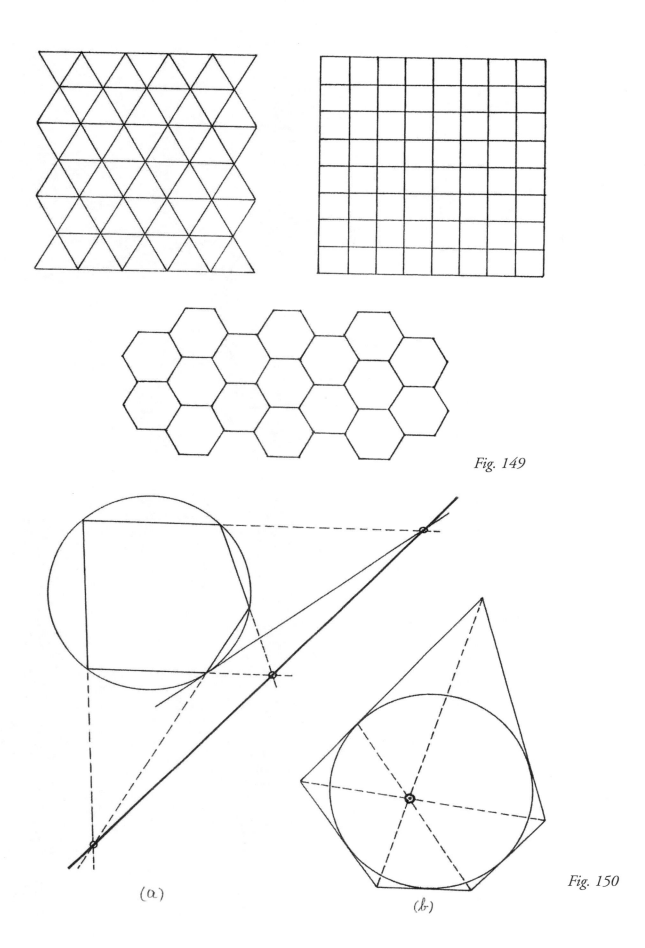

Fig. 149

(a) (b)

Fig. 150

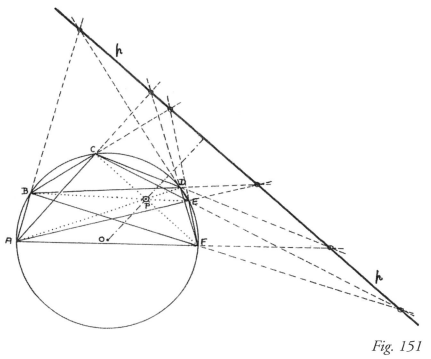

Fig. 151

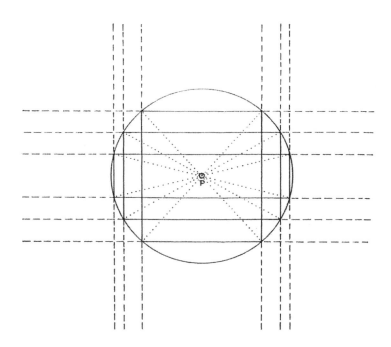

Fig. 152

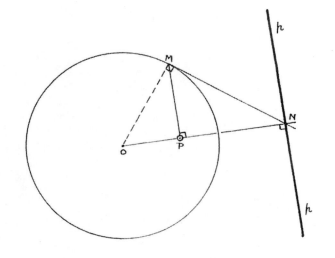

Fig. 153

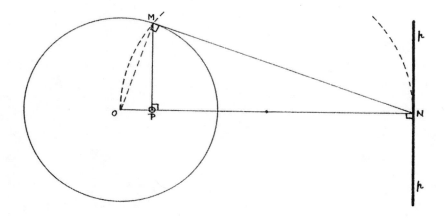

Fig. 154

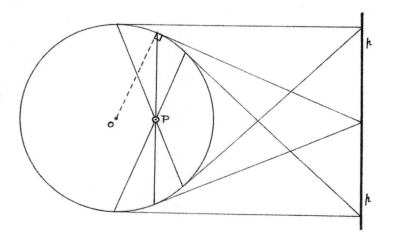

Fig. 155

Earlier in this chapter we said that lines ray through space, weaving the countless forms that occupy space. In Pascal's theorem, we have a striking example of this. Referring to Fig. 137, we see how a particular hexagon is formed by the weaving of six lines raying into the space of the circle in pairs from three points that lie in a straight line. Every hexagon in a conic is related to a particular straight line. Or, we may say that the hexagon is created from the straight line. Now when the line is at infinity, the hexagon may become regular (Fig. 138), the pairs of rays from the three points being parallel. This is a very significant fact in relation to form in nature. When nature chooses to create forms or cells with straight lines or edges, she generally chooses the regular hexagonal form. The reader is referred to D'Arcy W. Thompson's book *Growth and Form,* where a whole section of Chapter 7 on "The Forms of Tissues or Cell-Aggregates" is concerned with hexagonal symmetry. Here we will mention only one example from each of the kingdoms of nature. Perhaps one of the most beautiful and purest of the minerals is quartz (commonly called crystal), which crystallizes in hexagonal prisms capped with pyramids. In the plant kingdom, the lily is the traditional flower of purity. (One remembers the lily carried by the Archangel Gabriel in paintings of the Annunciation by the Old Masters.) And in the delicate epidermis of a leaf or young shoot of a monocotyledon, hexagonal cells of the most exquisite regularity are revealed. A creature of the animal kingdom that expresses, as a kind of nature parable, purity and selflessness in its whole activity is the bee, and the bee's cell or honeycomb is a most beautiful hexagonal structure. These are rather special examples, and as D'Arcy Thompson says, "The hexagonal pattern is illustrated among organisms in countless cases, but those in which the pattern is perfectly regular, by reason of perfect uniformity of force and perfect equality of the individual cells, are not so numerous. The hexagonal cells of the pigmented epithelium of the retina are a good example. . . . An equally symmetrical case, one of the first-known examples of an 'epithelium,' is to be found on the inner wall of the amnion, where, as Theodor Schwann remarked, 'Die sechseckigen Plättchen sind sehr schön und gross'."[21]

Various utilitarian reasons have been given for the prevalence of the regular hexagon as a fundamental form in nature. First, it is entirely economical in the partitioning of space; the only straight-line regular figures that will form a network filling a plane without any gaps or interstices are the equilateral triangle, the square, and the hexagon (Fig. 149). Of these, the strongest form is the hexagon. D'Arcy Thompson devotes some 20 pages of his book to consideration of the bee's cell. He quotes Pappus, the eminent Alexandrine geometrician of the third century A.D., as having come to the conclusion that bees were endowed with "a certain geometrical forethought" and that from the three possible forms, which of themselves can fill up the space round a point, "the bees have wisely selected for their structure that which contains most angles, suspecting indeed that it could hold more honey than either of the other two." Darwin, in *The Origin of Species,* refers to the bee's architecture as "the most wonderful of known instincts" and says that "beyond this stage of perfection in architecture natural selection could not lead; for the comb of the hive-bee, as far as we can see, is absolutely perfect in economizing wax."[22] Now according to D'Arcy Thompson, later research has shown that this "economy" theory is not correct. "The bee makes no economies; and whatever economies lie in the theoretical construction, the bee's handiwork is not fine nor accurate enough to take advantage of them." Again he says "that the beautiful regularity of the bee's architecture is due to some automatic play of the physical forces, and that it were fantastic to assume (with Pappus and Reaumur) that the bee intentionally seeks for a method of economizing wax, is certain; but the precise manner of this automatic action is not clear. . . . The question is, to what particular force are we to ascribe the plane surfaces and definite angles which define the sides of the cell in all these cases, and the ends of the cell where one row meets and opposes another?"[23]

Is not an answer to this question perhaps to be found in an understanding and interpretation of projective geometry? It is one of the objects of this chapter to show that in the study of this geometry, we are learning about the very nature of space itself and that the thought-forms required in this study lead us from the geometry itself out into different realms of knowledge and experience. Can we really ever understand the coming into being of, say, the bee's cell by considering "automatic action" within the sphere of physical matter and physical forces? Surely, projective geometry tells us to look elsewhere—to look out into the wide circumference of the universe and to recognize that although the earth gives the substance (wax, in the case of the bee's cell), the form of the hexagon arises through the activity of the bee working in harmony with cosmic forces pouring into the substance from the infinitudes of space. So, too, may we not conceive that the hexagonal form of the quartz crystal was built into the silica substance by such forces working onto the earth from the infinite periphery, but this time directly, not through the cooperation of any living organism? Thus the regular hexagon related to the Pascal line at infinity may indeed be considered not merely as a fact of geometry divorced from natural phenomena but as a thought-form that belongs intrinsically to such phenomena.

Is there anything corresponding to the Pascal and Brianchon theorems for a pentagon? Figs. 150a and 150b show a pentagon drawn inside and outside a circle. In Fig. 150a we see that the pentagon has two pairs of opposite sides. If we produce the remaining side and draw the tangent to the circle at the opposite corner, we get the third point, which lies in the same straight line as the other two. The tangent may be considered as the sixth side of a hexagon "shrunk to nothing." In other words, Pascal's theorem still holds for a pentagon when we treat it as a "degenerate hexagon." Treating the pentagon outside the circle in a similar manner, we see that Brianchon's theorem is applicable (Fig. 150b); the third line joins the remaining corner to the point of contact of the opposite side with the circle. This point of contact is the "degenerate" sixth corner of the figure considered as a hexagon. Corresponding drawings may also be done with respect to the other conic section curves.

Now we come to a figure with four sides—a quadrilateral or quadrangle,[24] and this brings us to the very important and well-known theorem of pole and polar. Fig. 151 illustrates this theorem. We see here three quadrangles inscribed in a circle—ABDE, ACDF, and BCEF—which are related to one another in that their diagonals all pass through a common point P. If we produce the opposite pairs of sides of these three figures, we get six points that all lie in a straight line (pp). The line is said to be the polar of the point P, which is called the pole. The number of quadrangles that may be drawn is not limited to three. Any number may be drawn so long as their diagonals all pass through the pole P. Therefore we can obtain an infinite number of points lying in the polar (pp), although many of these points will be "off the paper." It may be remarked here that the only difficulty in carrying out this and the previous drawings of this chapter is in getting them on the paper. This needs a little forethought, and then after the first drawing is done, one's imagination can picture the further possibilities. The only instruments required are a straight edge, a compass, and a sharp pencil; it may be emphasized again that no measurement of any kind is involved. It will be evident that the position of the polar line outside the circle is entirely dependent on the position of the pole within the circle and, furthermore, that the line joining the center of the circle O and the pole P is always at right angles to the polar (pp) (this is shown in chain line). If the pole is near the circumference, then the polar is near the circle; if the pole is near the center, then the polar is very far away. This leads us to consider the special case when the quadrangles are parallelograms, and as they are to be inscribed in a circle, they must at the same time be rectangles. Three of these are shown in Fig. 152, although there are an infinite number that can be drawn. The diagonals of all these rectangles pass through the center of the circle, which is therefore the pole P. Where is the polar line and in what direction? It is at infinity and is at one and the same time in all directions. In other words, what would seem to be an infinitely big circle may be regarded as a straight line, the so-called line at infinity. In this case then, pole and polar are related to one another as point center and infinite horizon.

Here again we have perhaps a picture of a universal phenomenon. The tiny unicellular organism can only be really understood as a "center" into which stream forces from the infinite circumference of the universe.

The theorem of pole and polar applies to all the conic section curves. For the circle, however, there is another construction that does not involve the drawing of quadrangles; this is illustrated in Fig. 153. P is the pole within the circle of center O; join OP, and at P draw a semi-chord at right angles to OP cutting the circle in M. Draw a tangent to the circle at M (at right angles to the radius OM), cutting OP produced in N. The line drawn at N at right angles to ON is then the polar (pp) of the pole P. This relationship of pole and polar is reciprocal; given the pole, we can find the polar; given the polar, we can find the pole. Fig. 154 shows the corresponding construction for this: the polar line pp is given, and from the center O of the circle, a perpendicular ON is dropped onto the line and a semicircle is drawn on ON as diameter, cutting the circle in M. From M, drop a perpendicular MP onto ON; then P is the required pole. Now the pole may be outside the circle; then the polar cuts the circle. For example, if in either of the two foregoing drawings, the point N is the pole, then the polar is the line MP (or rather MP is part of the polar). Thus when the pole is at infinity, the polar coincides with a diameter of the circle. Fig. 155 illustrates a further relationship: the pairs of tangents drawn at the extremities of the chords passing through the pole, P, meet in points lying on the polar.

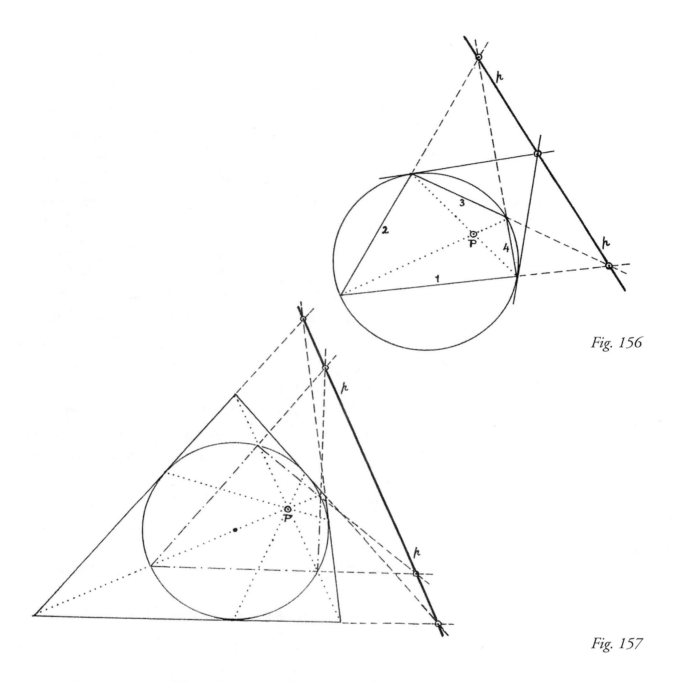

Fig. 156

Fig. 157

If, as in the case of the pentagon, we also consider the quadrangle as a degenerate hexagon, then we can treat it from the point of view of the Pascal and Brianchon theorems. Fig. 156 again shows a quadrangle inscribed in a circle. The opposite sides 1,3 and 2,4 are produced to give two points; the other two sides of the degenerate hexagon are the tangents to the circle at opposite corners of the quadrilateral, and these tangents meet in the third point on Pascal's line (the polar pp). It will be noticed that what we have described here is a combination of two of the methods of constructing pole and polar already described, the producing of opposite pairs of sides of the quadrilateral, and the drawing of tangents at the extremities of a chord through the pole P. (The tangents at the other pair of corners could be drawn to give the same result, only here they would not meet on the paper.) In Fig. 157 we have a quadrilateral circumscribing a circle; join the opposite pairs of corners and also the points where an opposite pair of sides touch the circle (these points being the two degenerate angles), and we get three lines lying in a Brianchon point, P. (A fourth line also passing through P may be drawn joining the points of contact of the other pair of sides.) Now the diagonals of the quadrilateral cut the circle in four points, which therefore determine an inscribed quadrilateral (drawn in chain line), whose opposite pairs of sides produced determine the polar (pp) of the pole P.[25] Thus from these considerations of the quadrilateral as a degenerate hexagon, the principle of duality applied to the theorem of pole and polar shows us that this theorem is self-dual.

We have now considered the remarkable relationships of lines and points in a plane in connection with six-sided, five-sided, and four-sided figures inscribed in or circumscribed to any of the conic section curves. It now remains for us to study a three-sided figure—a triangle—in a similar way.

Triangles of every shape may be drawn inside or outside a conic, so in this case there is no need to draw the conic. Fig. 158 shows a triangle and a point inside it. Rays are drawn from the angular points of the triangle through this point onto the opposite sides. This gives three points that determine a smaller triangle inscribed in the larger one. Now produce the opposite pairs of sides of these two triangles and we get three points that lie in a straight line. The position of the line and its distance from the triangle depend on the position of the point within the triangle. The line may be anywhere outside, as the point may be anywhere inside; we will therefore consider the case when the line is at infinity. This means that the three points that determine the line are at infinity and that each of these three points is determined by a pair of parallel lines; this will occur when the sides of the small triangle are parallel to the sides of the larger one (Fig. 159). Where then is the point within the triangle? Elementary geometry proves that it is the point where the medians intersect. (The medians are the three lines joining the angular points of a triangle to the midpoints of the opposite sides.) What is this point? It is none other than the center of gravity or center of balance of the triangle. If we cut a triangle out of a piece of cardboard, we shall find that this triangular lamina will balance on the point of a pin placed at this point. Thus the center of harmonious balance is related to the infinite circumference. We see that this is entirely in keeping with what we have found before: that perfection of form and harmony of balance on the earth come about through the dynamic activity of forces playing into physical substance from the infinite circumference of the universe. During the past 50 years, some leading scientists have become increasingly aware that such influences exist and that an understanding of them is essential for a true knowledge of the phenomena of the physical world. At the risk of being considered "unscientific" or "unmathematical," the author would suggest that what we have said here as an interpretation or application of these conceptions of projective geometry belongs also to man's relation to the spiritual world. Man can reach his full stature and possibilities only insofar as he relates himself to the world of spirit to which his real inner being belongs. He is a citizen of two worlds—the outer world of physical space and an inner world of spiritual experience. He belongs to the earth and to the heavens as point center does to the infinite circumference encircling it. If this is a true picture, then clearly, the concepts of infinity and of spiritual or divine are not unrelated. Is it only accidental that the word *infinite* is used in mathematics and also in theology? "The Infinite" is often used as a name for the Godhead.

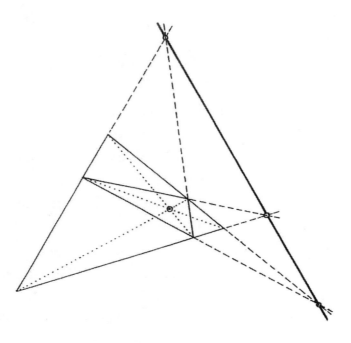

Fig. 158

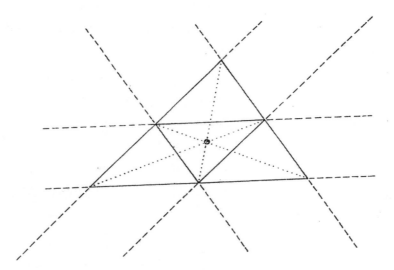

Fig. 159

Some years ago, there was a well-known religious book by the American, Ralph Waldo Trine, with the title *In Tune with the Infinite*. From the sphere of religion, this title expresses the ideas we have been endeavoring to develop out of the realm of mathematics. All true knowledge is a unity, and a truth in one sphere of human life and experience surely illuminates a truth in a different sphere. In fact, the truth may be one and the same, and we simply recognize different aspects of it much in the same way as when we look, say, at a tree, we perceive the reality of the tree only from one point-of-view.

Now, to return again to our geometrical considerations. The point and line obtained by the triangle drawing are in pole and polar relationship. Here, the mediating conic section curve would be an ellipse passing through the angular points of the smaller triangle and having the sides of the larger triangle as tangents, that is, an ellipse circumscribed to the smaller triangle and inscribed in the larger one. If the triangle is equilateral and the pole is at its center, then the ellipse becomes a circle. In such a particular case, the center of balance and the center of the circle circumscribed to the small triangle and inscribed in the larger one are coincident. We may also consider this theorem as we have done with previous ones as a degenerate case of either Pascal's or Brianchon's theorems. Actually, however, it is a particular case of one of the most fundamental theorems of projective geometry, to which the theorems of Pascal and Brianchon and pole and polar are closely related. This is known as the theorem of Desargues after the French mathematician of that name to whom we have already referred as one of the pioneers in this realm of geometry. Gerard Desargues was born at Lyons in 1593 and died in 1662; he was by profession an engineer and architect, but he gave courses of lectures on geometry in Paris from 1626 to 1630, which made a great impression on his contemporaries. In 1636 he published a work on perspective, and this was followed in 1639 by his researches in pure geometry in which he laid the foundations of what we now call projective geometry. His famous theorem may be stated as follow: If two triangles $A_1B_1C_1$ and $A_2B_2C_2$, lying in the same plane, are such that the straight lines A_1A_2, B_1B_2, C_1C_2 meet in the same point O, then the three points of intersection of the sides B_1C_1, B_2C_2 and C_1A_1, C_2A_2 and A_1B_1, A_2B_2 lie in a straight line (o,o). Figs. 160, 161, and 162 illustrate the theorem.

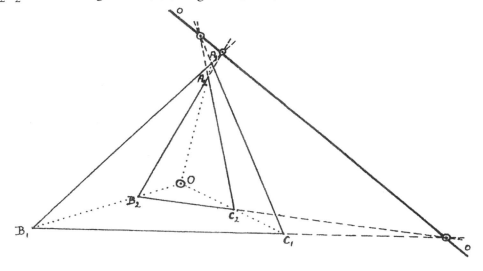

Fig. 160

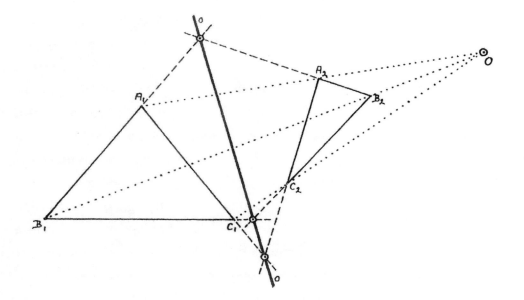

Fig. 161

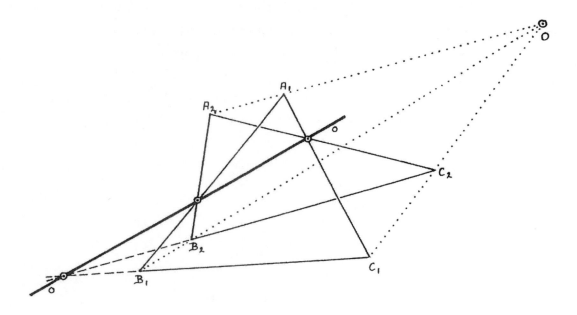

Fig. 162

206

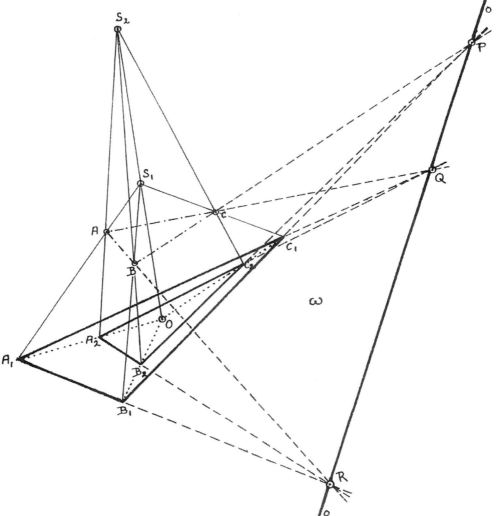

Fig. 163

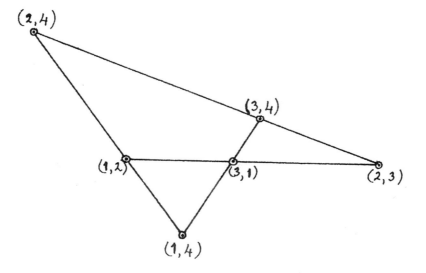

Fig. 164

207

The proof of Desargues' theorem can only be achieved by going outside the plane of the two triangles; that is, it is a three-dimensional proof, illustrated in Fig. 163. Through the point O, which is common to the straight lines A_1A_2, B_1B_2, C_1C_2, draw any straight line outside the plane ω, and in this straight line take two points S_1 and S_2. Project the triangle $A_1B_1C_1$ from S_1 and the triangle $A_2B_2C_2$ from S_2. The points A_1, A_2, O, S_1, S_2 lie in the same plane; therefore, S_1A_1 and S_2A_2 cut one another, say in A. Similarly, S_1B_1 and S_2B_2 cut one another, say in B and S_1C_1 and S_2C_2, say in C. The straight lines BC, B_1C_1, B_2C_2 intersect in pairs and therefore meet in one and the same point, P. (BC is the intersection of the planes $S_1B_1C_1$ and $S_2B_2C_2$, which do not coincide, so that the straight lines BC, B_1C_1, B_2C_2 do not all lie in one plane. The three planes BC.B_1C_1, BC.B_2C_2, and B_1C_1.B_2C_2 (or ω) intersect in the same point P.) Similarly, AC, A_1C_1, A_2C_2 meet in a point, Q, and AB, A_1B_1, A_2B_2 meet in a point, R. The three points P, Q, R lie in a straight line, which is common to the planes ω and ABC. The theorem is, therefore, proved, and the picturing of the proof expressed as it is in different planes is a fine exercise for the mathematical imagination.

We will now consider Desargues' theorem in a somewhat different manner. Let us imagine five points at random in space, each one joined to the other four by straight lines and then these lines produced to meet the plane of the paper. There will clearly be 10 such projecting lines or rays terminating in 10 points in the plane of the paper. These points will lie in threes in straight lines, and there will therefore be 10 straight lines. Thus we get a Desargues' configuration. One such configuration is shown in Fig. 163 by the points A_1, B_1, C_1, A_2, B_2, C_2, O, P, Q, R in the plane ω, which are formed by joining the points S_1, S_2, A, B, C (not in the plane ω) each to each of the others and producing these lines on to the plane ω. Now it is a very interesting exercise to build up the Desargues' configuration using only the plane of the paper. Such a configuration is shown in Fig. 165, and what follows is a description of this drawing. However, as it is somewhat involved we will consider first Fig. 164. We start with the horizontal line and take three points at random in it—points (1,2), (3,1), (2,3). The notation (1,2), and so on, indicates the point at which the line joining the points 1 and 2 (two of the five points in space outside the plane of the paper) meets the plane of the paper. Thus the lines joining the points 1, 2, and 3 meet the plane of the paper in the points (1,2) (3,1) (2,3), and these three points must of course lie in a straight line. We next bring into consideration point 4, and therefore take outside the first line any point (1,4) and join this to point (1,2) and produce to any third point on this line, which will be (2,4). These three points (1,4) (1,2) (2,4) are where the lines joining the 1, 2, and 4 meet the plane of the paper. Next join points (2,4) and (2,3); then join points (1,4) and (3,1). Somewhere along this last line must be the point (3,4); but the point (3,4) must also be on the line joining (1,4) and (3,1), and therefore its position is fixed and determined unlike the other five points. This configuration of Fig. 164 is therefore the result of joining each of four points in space (no three of them being in the same straight line) to each of the others. Producing the joining lines onto the plane of the paper, we see that it is a quadrangle—four lines and six points. Now, finally, we consider point 5 in relation to the other four points (see Fig. 165). Point (1,5), say, is chosen at random; point (1,4) is joined to (1,5) and the line produced to any point (4,5); join (2,4) and (4,5) and produce to (2,5), which is on the same straight line as (1,2) and (1,5); thus point (2,5) is determined. Join (2,5) and (2,3) and (1,5) and (3,1) and produce to meet in (3,5), which is therefore also determined; then we find that the remaining trio of points (4,5) (3,4) (3,5) lie in a straight line (shown as a chain line). In Fig. 165, we see that this Desargues' configuration consists of 10 points and 10 lines; this is confirmed mathematically by the fact that there are 10 possible combinations of 5 numbers taken in pairs:

$$C_2 = \frac{5 \times 4}{2 \times 1} = 10.$$

Three of the points lie on each of the lines, and three lines pass through each point. Thus we again have a self-dual figure.

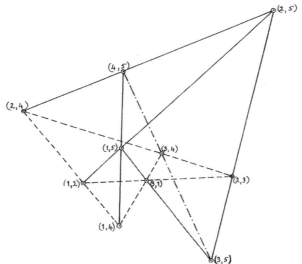

Fig. 165

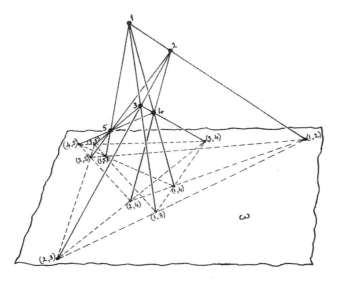

Fig. 166

Much of what we have said about the theorem of Desargues, especially its proof in three dimensions, requires quite a considerable capacity for spatial imagination. As a kind of confirmation of what we have arrived at by our thinking and imagination, a mathematical model can be very helpful and instructive. The making of such a model also requires skill and ingenuity, and there will always be the boy or girl for whom such an activity gives great satisfaction. Fig. 166 is a picture of such a model. One may make small spheres (about ¾ inch in diameter) out of colored wax to represent the five points in space, and these may be fixed by lengths of aluminum knitting needle stuck into a cork that is held rigidly by a framework attached to the edges of the drawing board—that is, the plane, ω, onto which the projection is to be made. ("Meccano" is ideal for the framework.) Then the lines joining the five points and produced onto the drawing board may be represented by long knitting needles, and where they meet the board (covered with a sheet of drawing paper), they may be fixed into position on the paper with a small piece of plasticine stuck onto the paper with cellophane. The plasticine points are then joined in three's as shown (broken lines in Fig. 166), and we have a Desargues' configuration. (The picture of the model does not show the cork holding the five wax spheres nor the framework supporting it.) It should be mentioned that the choice of position of the five small spheres will be somewhat of a problem, so make sure that the projecting knitting needles meet the drawing board. (We have, of course, already drawn a picture of such a model in Fig. 163.)

As we have seen, the Desargues' configuration has 10 points and 10 lines; the 10 drawings of Fig. 167 are all the same configuration (a plane view of the configuration in the model picture), showing every possibility of pairs of triangles in the Desargues' relationship. What a wonderful relationship this is of lines and points in a plane—10 different cases of Desargues' theorem all in the one configuration! Here indeed is a kind of "magic" of space.

Another very important theorem of projective geometry that again reveals the relationship of lines and points in a plane is that of the harmonic quadrangle leading to the conception of harmonic points. In a straight line, a pair of points A and B is chosen at random (Fig. 168). A pair of rays is drawn from each of these points to form a quadrangular cell (shown shaded), and then the diagonals of the cell are drawn to meet the straight line in P and Q. Now another pair of rays is drawn from A (shown below the line) and a ray from P cutting them. Through these two cutting points, a pair of rays is drawn into B forming a second quadrangular cell. Lastly, draw the other diagonal of this second cell, and it will be found to fall on the line exactly at the point Q. The two cells thus formed are called harmonic quadrangles, and the pairs of points A,B and P,Q are said to be harmonic points or harmonic conjugate pairs. The cells are drawn on either side of the line for the sake of clarity of construction.

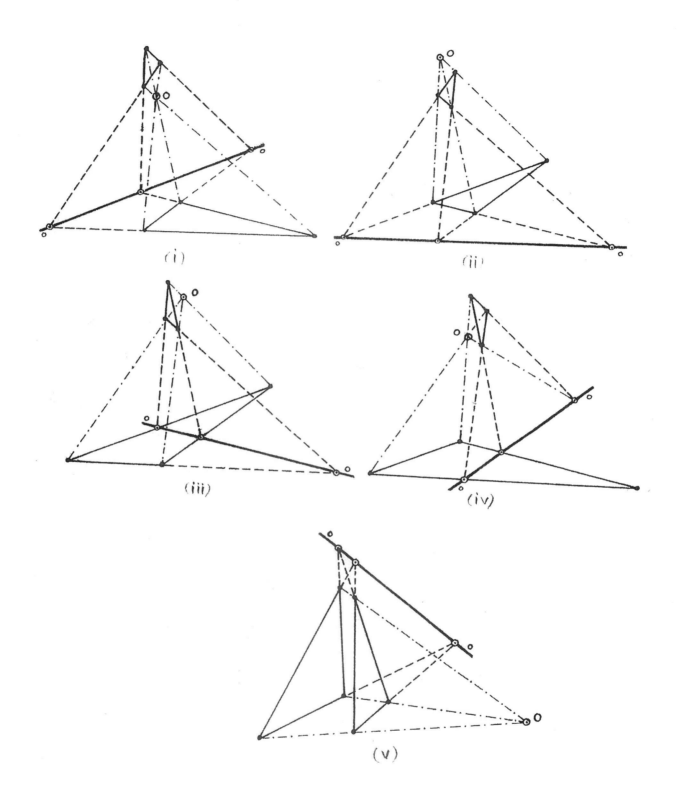

Fig. 167 i–v

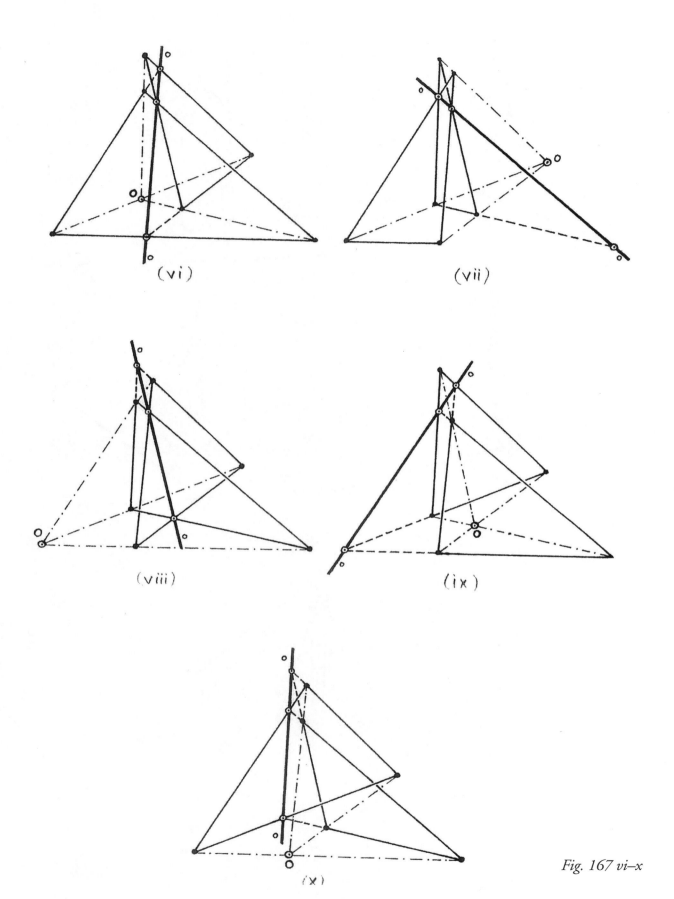

Fig. 167 vi–x

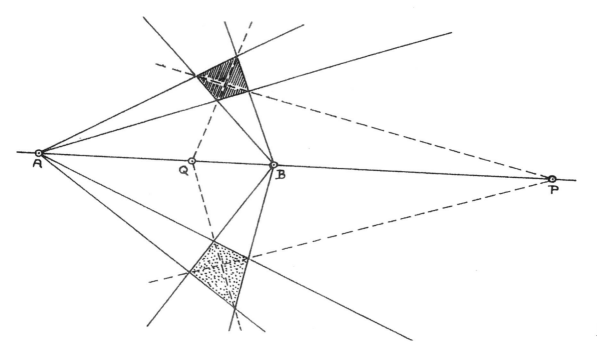

Fig. 168

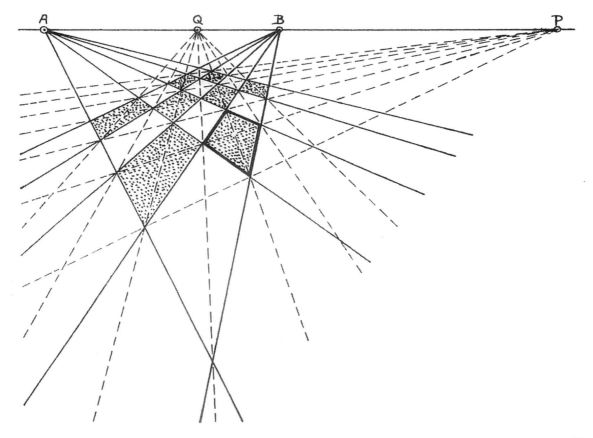

Fig. 169

Fig. 169 shows a family of harmonic quadrangles drawn below the line. The cell originally drawn is shown in thick line, and the others arise quite naturally from this cell by the above construction. We see at once that this is a perspective drawing. This becomes even more evident if we arrange the harmonic conjugate pairs of points a little differently as in Fig. 170. It will be clear from the construction that if a pair of points (A,B) is chosen and a third point (P) in the same straight line, then a fourth point Q in the line is determined. Let us suppose that points A and B remain fixed. Then for every position of point P, there is a definite position of point Q. As P moves along the line from B to the right, Q moves from B to the left, and the further P goes, the nearer Q approaches the midpoint between A and B. When P is at infinity, Q is exactly halfway between A and B (Fig. 171). Then as P returns from infinity and approaches A more and more nearly (still moving in the same direction along the line from left to right), Q gets closer and closer to A (still moving from right to left). Thus as P moves from B to A via infinity, Q moves from B to A by the "short route." This construction of Fig. 171 gives a very quick and easy way of bisecting the distance between two given points without using compasses. Q is a balancing point, a fulcrum if we think of a balance for weighing, between A and B. It is just this balancing point that is related to a point at infinity, and we see once again how the infinite plays into the world of physical space to create perfection of form, balance, and harmony. Fig. 172 is a simple perspective drawing of a chessboard placed with two of its sides parallel to the horizon line and viewed straight down the middle. Comparing this with Fig. 171, we see that the pair of points A and B are here the two "distance points" (D.P.), and the point Q is the "center of vision" (C.V.). The point P is at infinity and hence the lines parallel to the horizon line.

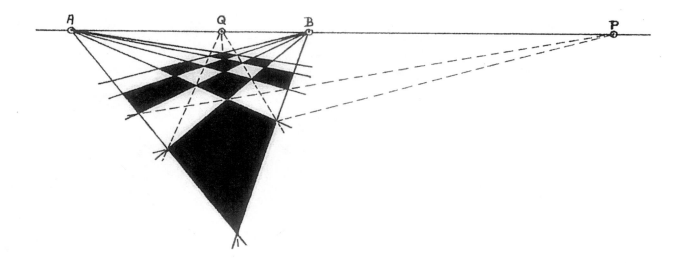

Fig. 170

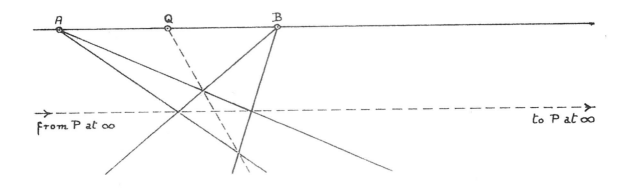

Fig. 171

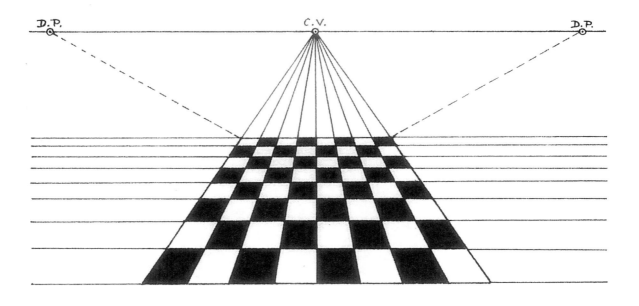

Fig. 172

In a previous drawing, we have actually come across harmonic points. It is the pole and polar construction with respect to a circle. Fig. 173 again illustrates this construction. The vertical line through P is the polar of the pole Q, and P and Q are conjugate harmonic points with respect to a second pair, A and B (and these latter points are determined by the circumference of the circle cutting the line PQ). That this is so is shown by the fact that a harmonic quadrangle can be constructed with respect to these two pairs of points A,B and P,Q.[26] Here again, we have a projective relationship, and therefore, harmonic points will arise in connection with pole and polar in respect of any conic section curve.

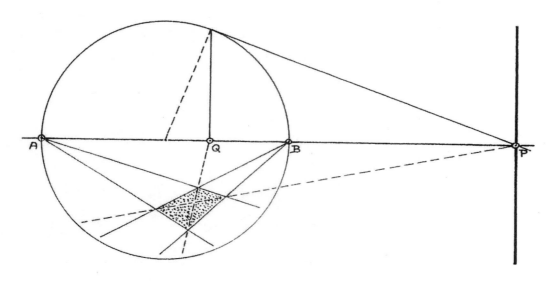

Fig. 173

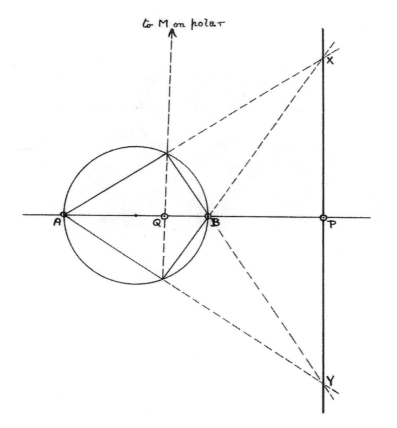

Fig. 173a

On the other hand, we can discover Desargues' theorem within the harmonic quadrangle construction. Fig. 174 begins with a repetition of Fig. 169, in that we have constructed a pair of harmonic quadrangles KLMN and K'L'M'N'. Now each of these quadrangles with its diagonals contains four triangles, for example, KLM, LMN, MNK, and NKL in the upper quadrangle. We will consider a pair of corresponding triangles, one from each quadrangle, say NKL and N'K'L'. Join the corresponding angular points of these two triangles (i.e., N'N, K'K and L'L), and we get three lines that lie in a point (O). Now produce the corresponding sides (i.e., KN, K'N'; KL, K'L'; NL, N'L'), and we get three points (A, B, and P) that lie in a straight line, and these points are three of the four harmonic points. This is, of course, the theorem of Desargues, which is to be found expressed four times in all within the figure because there are four pairs of corresponding triangles. The point O is the same for all four, but the three points in a straight line will be A, B, and P for the pairs of triangles NKL, N'K'L' or NML, N'M'L', and A, B, and Q for the pairs of triangles KLM, K'L'M' or KNM, K'N'M'.

This relationship of harmonic quadrangles and harmonic points to the theorems of pole and polar and of Desargues again shows us what remarkable connections there are among the various theorems of projective geometry.

Just as a pair of points and a third point in a line determine a harmonic fourth point, so do a pair of lines and a third line in a point (and in a plane) determine a harmonic fourth line. Here again, we have the principle of duality, and we will express the two constructions side-by-side and notice that they are identical except for the interchanging of the words *line* for *point* and *point* for *line*, as well as two or three relevant words to make sense, for example, *pass through* is changed to *lie in*.

Harmonic Points in a Line (Fig. 175)	Harmonic Lines in a Point (Fig. 176)
Given a pair of points A,B and a third point P in any line of a plane, we can draw any number of quadrangles (KLMN) in the plane, so that five of their six lines pass two-by-two through A and B and one through P. Then their sixth line will always pass through one and the same point Q of the originally given line, and this is the point harmonically paired with P with respect to A and B.	Given a pair of lines a,b and a third line p in any point of a plane, we can draw any number of quadrilaterals (KLMN) in the plane, so that five of their six points lie two-by-two in a and b and one in p. Then their sixth point will always lie in one and the same line q of the originally given point, and this is the line harmonically paired with p with respect to a and b.

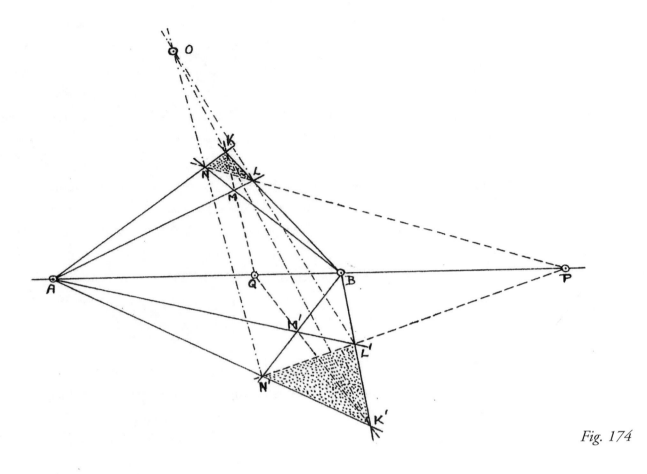

Fig. 174

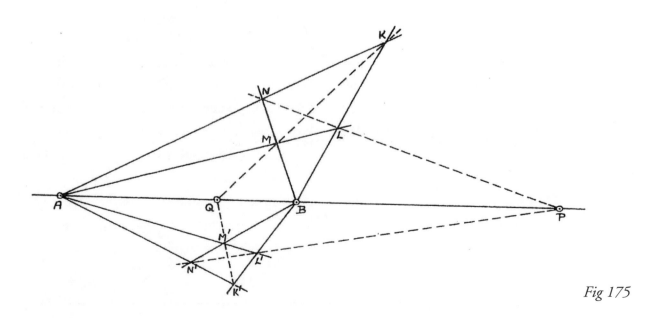

Fig 175

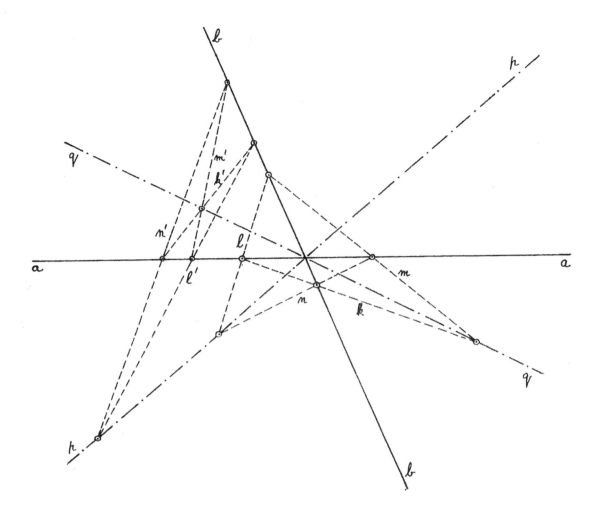

Fig. 176

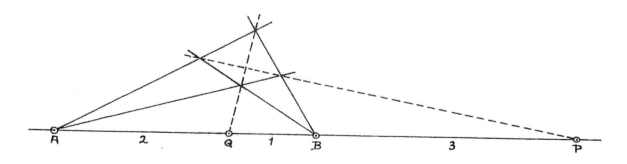

Fig. 177

219

In our studies of projective geometry so far, all that we have done has been quite devoid of any thought of measurement. Now these important and far-reaching conceptions of harmonic range of points in a line and harmonic range of lines in a point lead us to consider metrical properties, not arbitrarily, but arising out of the very nature of the relation of lines and points in a plane. We have already come across such a metrical property in the special case of the harmonic quadrangle where the point P is at infinity (Fig. 171). Here we have seen that the point Q is midway between the points A and B. That is, by this construction we have bisected the distance AB, or in other words, we have halved AB. If AB is considered of unit length, then $AQ = QB = 1/2$ unit length, or if AQ is considered of unit length then AB = 2 units.

In the above special and simple case, there is an exact relation between four harmonic points along a line, but this is not only true for a special case. The harmonic pair P and Q always divide the distance between A and B internally and externally in the same proportion (see Fig. 175):

Q divides distance AB internally, giving the ratio AQ : QB
P divides distance AB externally, giving the ratio AP : PB

(In carrying out the "external" division we always have to reverse direction; that is, PB is negative, by the usual sign convention.) Then

$$(AQ : QB) = - (AP : PB)$$

It should be noticed that this ratio is always >1 when P is to the right (Q is then nearer to B than to A) and is always <1 when P is to the left (Q is then nearer to A than to B), and the ratio will be nearer to 1 the further away P is in either direction. The ratio = 1 when P is at infinity.

Fig. 177 shows a harmonic range in which the ratio is

$$2 : 1 : -AQ : QB = 2 : 1 : AP : PB = 6 : 3 = 2 : 1$$

(neglecting the negative sign). Now the dual concepts of harmonic points in a line and harmonic lines in a point (Figs. 175 and 176) are clearly interrelated and may be illustrated in one and the same figure. In Fig. 178, the harmonic lines of Fig. 176 are reproduced, and any line, o, is drawn, cutting them in the points A, B, P, and Q. By the drawing of the harmonic quadrangle, we see that these points form a harmonic range in the line o.

The converse of the above relationship is, of course, also true. That is, if we start with a harmonic range of four points and confront them with a point O—an eye—outside the line, then the four rays from this point to the four harmonic points will be harmonic (Fig. 179). This harmonic property can be projected from one line or point to another as often as we please, and the harmonic relationship remains unimpaired. For example, any other line (*ll*) cutting the four rays in Fig. 179 gives a harmonic range of points (A',B'; P',Q') (see Fig. 190).

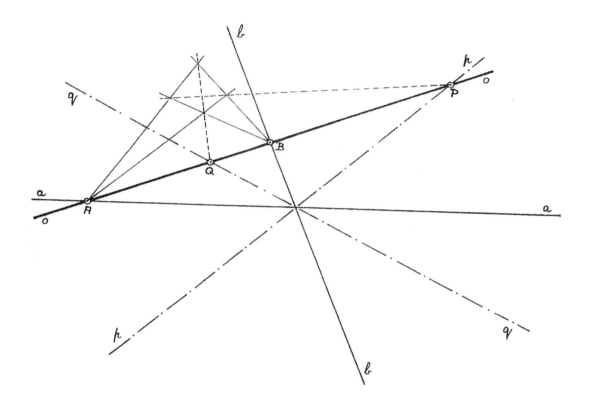

Fig. 178

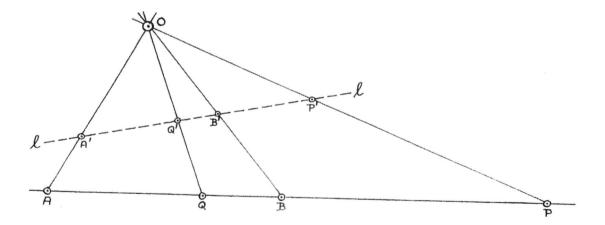

Fig. 179

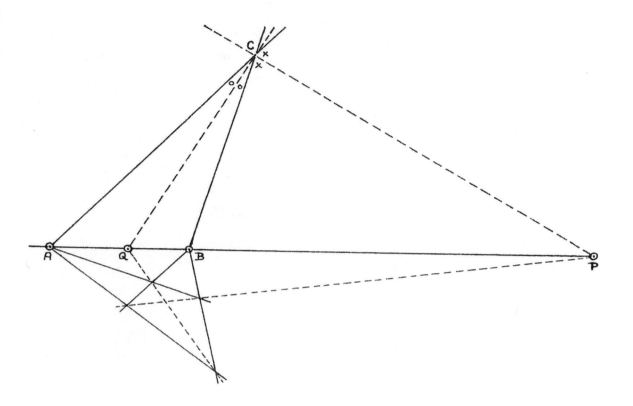

Fig. 180a

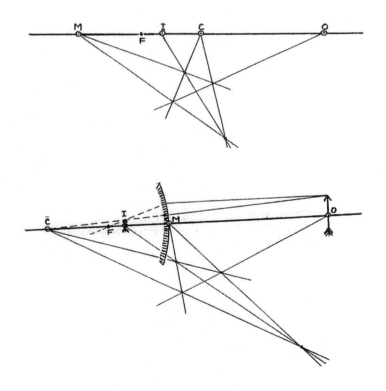

Fig. 180 b, c

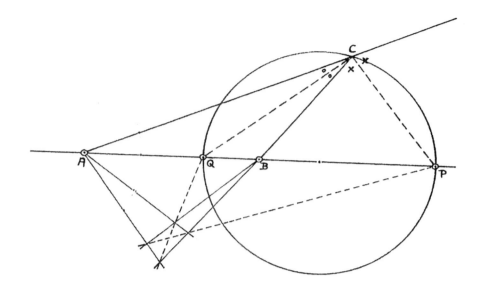

Fig. 181

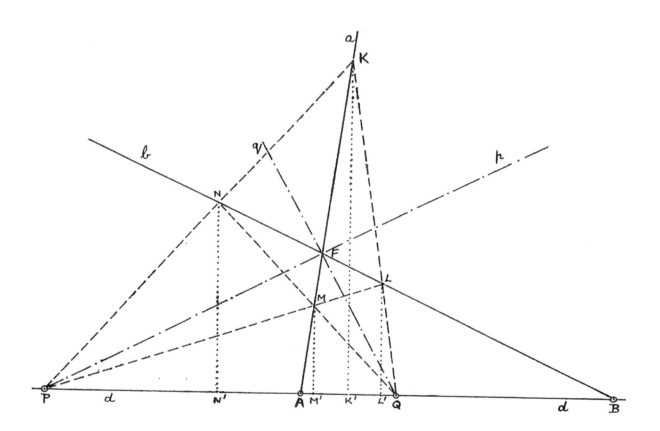

Fig. 182

Now with four harmonic lines in a point there are also relationships of measure—this time among the angles between the lines. As an example of this, we may refer to Fig. 180a. Here the bisectors of an angle of a triangle cut the opposite sides in points that separate harmonically the vertices on that side. It is a well-known fact of elementary geometry that the angle PCQ is a right angle. Thus the measure of a right angle is associated with harmonic conjugate points.

If we compare this drawing with that of Fig. 176 we see that they are really the same except that in this case one pair of harmonic lines is at right angles, and they bisect the angles between the other pair. This configuration of harmonic lines is related to the symmetry of rectangle and rhombus, which the children illustrated in their first geometry lessons. We refer to Fig. 36, where we see that the diagonals of the family of rhombuses are a pair of harmonic lines at right angles, bisecting the angles between the second pair, which are the diagonals of the family of rectangles.

This connection of an harmonic range of points with the measure of a right angle leads us to the Apollonian circles, which we considered in Chapter 5 as the geometrical picture of the process of division, that is, as curves expressing a constant ratio. Fig. 181 shows one such circle of Apollonius having a constant ratio of 2:1 and 1:2; that is, if C is any point on the circle, then $AC/BC = 2/1$. The focal points A and B are a conjugate pair with P and Q, the extremities of the diameter of the circle, as the other pair. That is, the points A, B and P, Q are harmonic points in a line and furthermore the harmonic ratio:

$$AQ : QB = AP : PB = 2 : 1.$$

We also notice that CQ and CP are the bisectors of the interior and exterior angles at C because the angle QPC is a right angle (an angle in a semi-circle).

It will already have become evident to the reader that the relationship of harmonic conjugate points plays a vital part in projective geometry. Indeed, it is one of the most fundamental theorems and will appear again in what follows concerning the conic section curves. Before proceeding to this aspect of our studies, we will consider a very important physical application. All phenomena of symmetry and reflection are related to the conception of harmonic conjugate points. So, even the simple symmetry drawings of little children have a real connection with the basic ideas of projective geometry. Let us consider one or two examples illustrating what we have indicated here. Everyone is familiar with the distance relationships of image and object in a plane mirror; the image is the same distance behind the mirror as the object is in front. This physical fact may be expressed by saying that object and image positions give a harmonic conjugate pair with respect to the mirror position and the point at infinity on the axis as the other pair. (See Fig. 171, in which Q is the position of the mirror, and A and B are the positions of object and image.) Such a reflection is, of course, also illustrated in the symmetry drawings.

What about reflection in a spherical mirror? We have considered this in another connection in the series of drawings (Fig. 128). Fig. 180b is a repetition of Fig. 128a showing the relative positions of object and image for a concave spherical mirror (M being the position of the mirror). By the harmonic quadrangle construction, we see at once that object and image are again a harmonic conjugate pair with respect to the mirror and its center of curvature as the other pair. (The plane mirror relationship considered above is really a special case since the plane mirror has its "center of curvature" at infinity.) If the object is at infinity—for example, a parallel pencil of light falling on the mirror—then the image is at the focus F, midway between the mirror and its center of curvature (see Fig. 171). Fig. 180c shows the construction for the image of an object in a convex mirror, and again the harmonic quadrangle construction reveals object and image as a harmonic conjugate pair with respect to the mirror and its center of curvature.

We may indeed say that the principle of harmonic conjugate points is the archetype for all processes of reflection and considerations of symmetry.

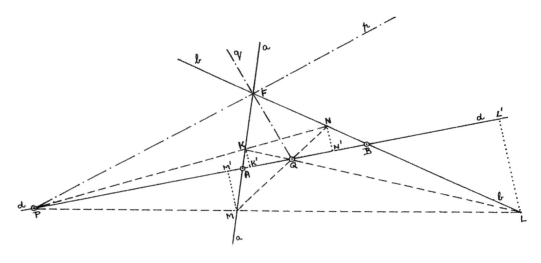

Fig. 183

We have seen in Chapter 7, and again in the present chapter, that the three curves of the conic sections—the ellipse, the parabola, and the hyperbola—are related to one another by principles of perspective and projection. In the particular case of four harmonic lines where two are at right angles bisecting the angles between the other two, we are again led naturally and directly to the conic section curves. In Fig. 182 four harmonic lines are drawn (a, b, p, q) so that one pair (p,q) is at right angles and therefore bisect the angles between the other pair. These harmonic lines give rise to a range of four harmonic points (A,B,P,Q) along the line d. These harmonic points in turn give rise to an harmonic quadrangle KLMN with its points lying in pairs on the pair of harmonic lines a and b. We shall now show that these four points K, L, M, N lie on a conic section curve. To do this, we have to prove that the ratio of the distances of each point from the point F and from the line d is a constant, that is, that

$$KF : KK' = LF : LL' = MF : MM' = NF : MN'.$$

(See definitions of parabola, ellipse, and hyperbola, Chapter 7.)

Triangle KFN has its exterior angle at F bisected by line FP.

Therefore: KF : NF = KP : NP (Euclid VI.3)

From similar triangles KPK', NPN' ; KP : NP = KK' : NN'

Thus, KF : NF = KK' : NN'

That is, KF : KK' = NF : NN'

Similarly, by considering the other three triangles NFM, MFL, and LFK we find that

$$NF : NN' = MF : MM'$$
$$MF : MM' = LF : LL' \text{ and}$$
$$LF : LL' = KF : KK'$$

Thus, KF : KK' = LF : LL' = MF : MM' = NF : NN'

and therefore the four points K, L, M, N lie on a conic section curve.

In the particular case of Fig. 182, the points K, L, M, N are nearer to the point F (the focus) than they are to the line d (the directrix); the constant ratio proved above is therefore <1, and thus the points determine an ellipse.

In Fig. 183, we have the case where the distances of the points from the focus F are greater than their distances from the directrix d; that is, the constant ratio of their distances = 1, and therefore they lie on a parabola.

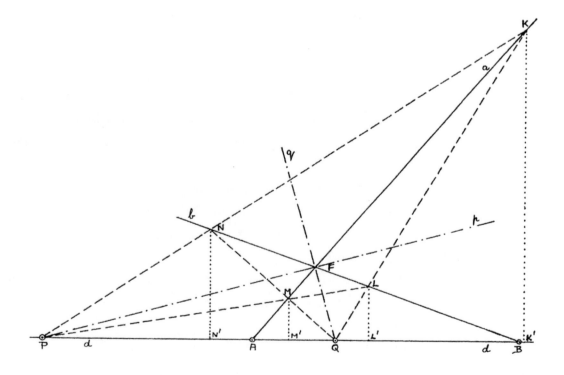

Fig. 184

To sum up then, the points K, L, M, N in each of the drawings of Figs. 182, 183, and 184 determine a conic section curve. In Fig. 182 they determine an ellipse because KF : KK' = LF : LL' = MF : MM' = NF : NN' = a constant < 1. This constant is called the eccentricity of the curve. In Fig. 183 the four points determine as hyperbola because the eccentricity constant is > 1, and in Fig. 184 they determine a parabola since the eccentricity = 1.

We will now construct the three curves applying the principle of harmonic points but also bringing to our aid the Apollonian circle, which is the curve of constant ratio from two fixed points. Our problem is to trace the locus of a point that moves so that its distances from a fixed point and a fixed line shall give a constant ratio. First then, the ellipse (Fig. 185): Take a fixed point F (the focus) and a fixed line d (the directrix) and choose a certain ratio < 1, say 2 : 3. We now have to find all points whose distances from focus F are to their distances from directrix d in the proportion of 2:3. Two of these points are A_1 and A_2, which divide the line FX, perpendicular to the directrix, internally and externally in the proportion of 2:3. This means, of course, that the four points F, X and A_1, A_2 are harmonic conjugate points. Through A_1 and A_2, draw lines parallel to the directrix. Now draw any ray through F cutting the directrix in Y and the two parallels in B_1 and B_2. Then from consideration of similar triangles we see that F, Y and B_1, B_2 are also harmonic pairs; that is, $FB_1 : B_1Y = FB_2 : B_2Y = 2$. If we now draw an Apollonian circle on B_1B_2 as diameter, this circle will contain all points whose distances from the two points F and Y are in the same proportion, 2:3. Finally, from Y draw a perpendicular to the directrix cutting the circle in P_1 and P_2. Then P_1 and P_2 are two more points that lie on the ellipse because $P_1F : P_1Y = P_2F : P_2Y = 2 : 3$. ($P_1Y$ and P_2Y are clearly the distances of P_1 and P_2 from the directrix).

By drawing many such circles for many rays through F (their centers will all lie in the vertical broken line, which is the right bisector of A_1A_2), we may obtain as many points of the ellipse as we please, and through these the curve may be drawn. We shall also find that all the circles will touch the ellipse and envelop it and that the ellipse can never go outside any of the circles. Thus if we draw a sufficient number of these circles, together they will mold the ellipse from the outside; that is, the curve of the ellipse will appear as a boundary of space within all the circles (Fig. 186).

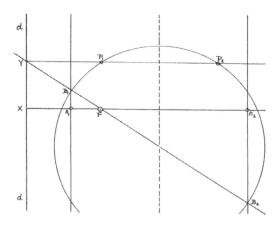

Fig. 185

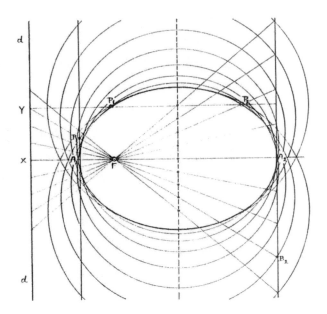

Fig. 186

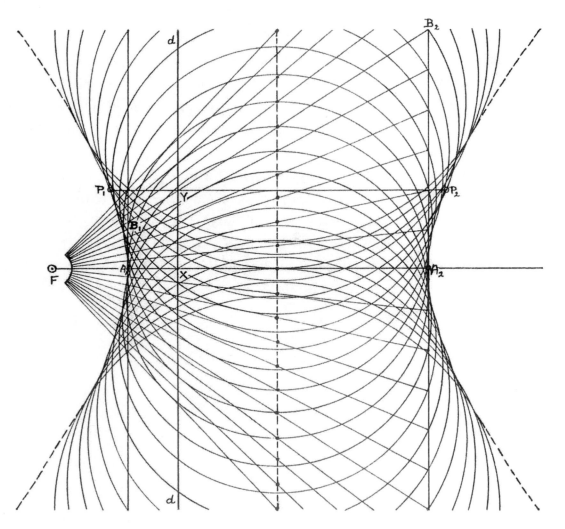

Fig. 187

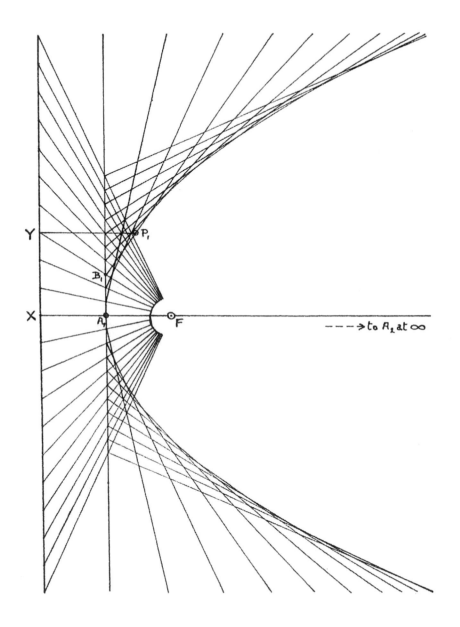

Fig. 188

To create a hyperbola, we repeat the above construction, only this time the eccentricity ratio must be >1; we choose the ratio 3:2. Now we see that the circles no longer "embrace" the curve, but they nevertheless still mold it from the outside. That is, the curve of the hyperbola appears as a boundary of space outside all the circles, although the circles are really, of course, outside the hyperbola (see Fig. 187).

Fig. 188 shows the same construction for forming a parabola. Here the point A2 is at infinity, and the circles drawn upon infinite diameters become straight lines at right angles to these diameters. The eccentricity ratio = 1. Once again, the curve—this time a parabola—is molded from the outside but now by straight lines, that is, by circles of infinity diameter.

The forming of these curves has arisen out of the consideration of the special properties of the harmonic quadrangle KLMN (see Fig. 182) with its right angles "diagonal triangle" (as it is called) PFQ. We will now therefore start from a given conic (an ellipse) with focus F and directrix d and derive the harmonic quadrangle and diagonal triangle (Fig. 189). Any quadrangle KLMN is drawn with its four points on the curve and two of its lines KM and LN meeting in the focus F. Then the other pairs of sides KN, LM and KL, NM must meet in points P and Q respectively on the directrix; the diagonal triangle PFQ has the directrix for one of its sides and the focus for its opposite corner, and moreover, it is right-angled at the focus F. If the lines KM and NL cut the directrix in A and B respectively, then the four points P, A, Q, B form a harmonic range along the directrix. We may sum this up by the statement that *any quadrangle of points on a conic section curve, having one of its three diagonal points at the focus, has its other two on the corresponding directrix, and the diagonal triangle made of all three points is right-angled at the focus.*

The focus and the directrix are, of course, always pole and polar with respect to the conic. Only when we are considering the focus as the pole is the diagonal triangle right-angled. Fig. 189a illustrates a more general case of pole and polar, where the diagonal triangle is not right-angled. Here we have a quadrangle ABCD inscribed in a circle, and the diagonals AC, BD give the pole P, and the opposite pairs of sides produced give two points Q and R, which determine the polar. Thus, PQR is the diagonal triangle. Now there is a further important relationship, which is shown in this figure: the tangents at A, B, C, D are drawn, giving a quadrilateral a b c d circumscribed to the circle, and this quadrilateral also determines the same pole and polar and the same diagonal triangle. This may be summed up as follows: Given a four-point figure ABCD, the points lying on a conic, and a four-line figure a b c d, the lines being tangents to the conic at the points A, B, C, D, then these two figures have the same diagonal triangle. This diagonal triangle may be considered as formed either of the three points DA.BC, DB.CA, DC.AB, or of the three lines da.bc, db.ca, dc.ab. (The point DA.BC is the point where the lines DA and BC intersect; the line da.bc is the line joining the point where the lines d and a intersect to the point where the lines b and c intersect.)

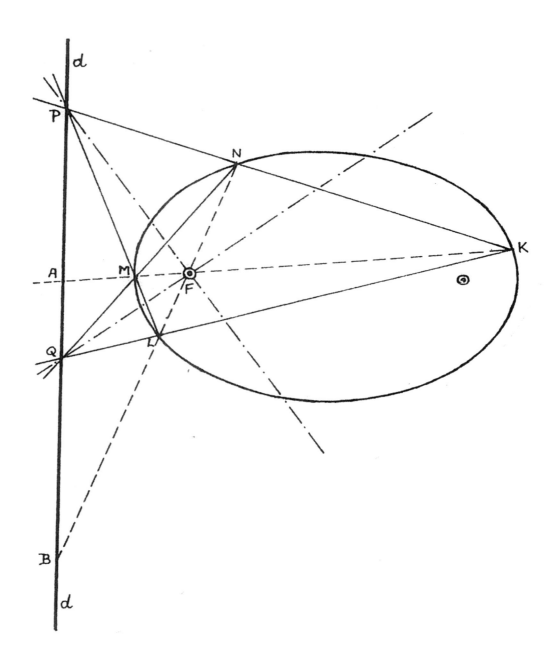

Fig. 189

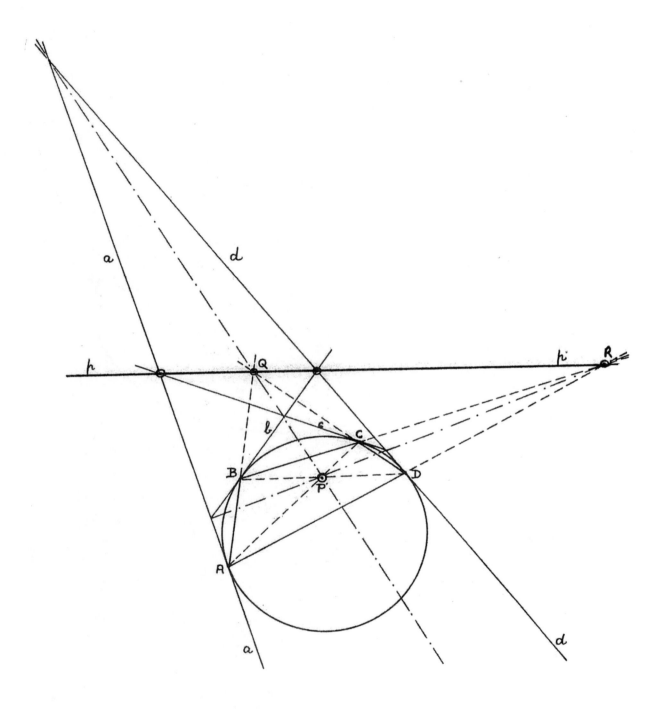

Fig. 189a

Fig. 189b is the particular case for a circle where the pole is at its center. The inscribed quadrangle is then a rectangle, the circumscribed quadrilateral is a rhombus, the polar is at infinity, and the diagonal triangle is again right-angled at the pole P with one of its sides in the infinite periphery. This figure illustrates the above statement in italics for the case where the conic section curve is a circle.

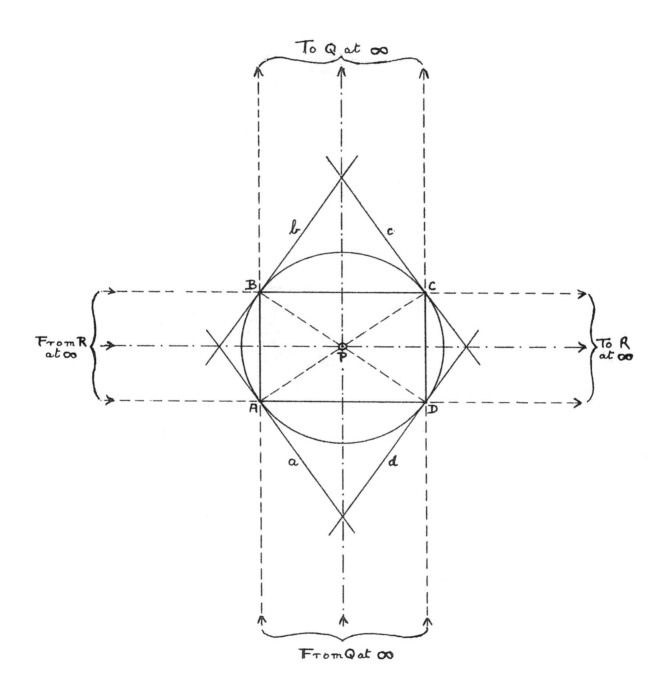

Fig. 189b

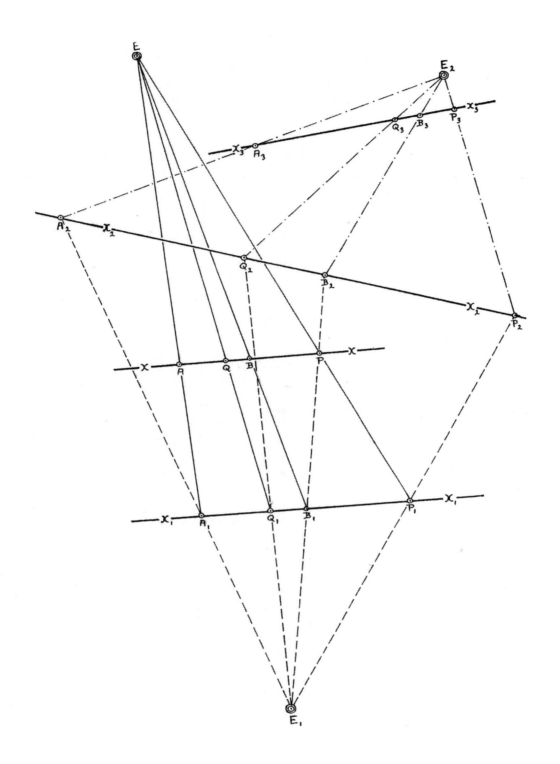

Fig. 190

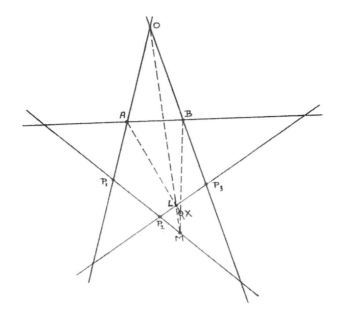

Fig. 191

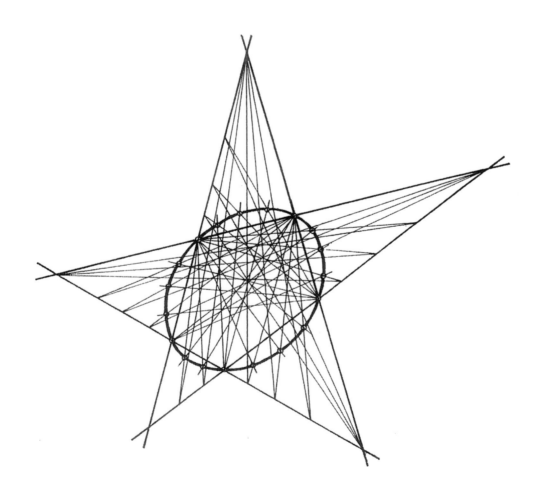

Fig. 192

We have seen in the last few pages how the study of the conic section curves belongs essentially to projective geometry because these curves arise quite naturally from the relationships of lines and points in a plane. They are inherent in the very nature of space, and their laws and properties are those of space itself. The construction we have used for the conics depends on four points in a plane arranged in such a manner that they give rise to a harmonic quadrangle having a right-angled diagonal triangle, the right angle being at the focus of the conic. Now if we add another point to the four, the statement about the construction of a conic is greatly simplified, for now we can say that *any five points in a plane determine a conic section curve so long as no three of them are in a straight line.* This is a very remarkable property: that through five points placed quite at random in a plane, an ellipse or a parabola or a hyperbola can be drawn. Such a construction of a conic is essentially a projective one, and before we embark upon it we will consider certain aspects of projection and perspective.

Suppose we have, say, four points along a straight line, and, as we have been dealing with harmonic ranges, we will arrange these four points in harmonic pairs (although such an arrangement is not essential to what we wish to show). Fig. 190 then gives such an ordering of four points A, Q, B, P, in a line x, such that AQ:QB = AP:PB = 2:1. If we now view these points from any eye-point E and project them onto another line x, parallel to the first one, then we get another harmonic range A_1, Q_1, B_1, P_1, which has the same ratio of 2:1 as the original one. This is clear from a consideration of similar triangles. Now we view this second range from another random eye-point E_1 and project it on to another line x_2 drawn obliquely to the direction of the other two, giving points A_2, Q_2, B_2, P_2. These points are also a harmonic range though with a different ratio from the other two. A third eye-point E_2 projects the points in line x_2 into another oblique line x_3 giving a further harmonic range A_3, Q_3, B_3, P_3 with again a different ratio from the previous ones. We see from this that however many projections we may make, the principle of the harmonic range cannot be destroyed, although the ratio of the harmony can be changed with every projection (see Fig. 179).

We shall now construct the conic section curves using five points placed at random and applying a sequence of projections. Fig. 191 illustrates the construction. The five given points are A, B, P_1, P_2, P_3, and they are joined in pairs as shown giving (in this case) a pentagram. Two of the points A and B are chosen as eye-points or "raying-out" points, and the other three P_1, P_2, and P_3 serve for fixing the rays. Rays from A give points (e.g., point L) in line P_2P_3; these points are viewed from eye-point O and projected into line P_1P_2 (e.g., point M). These points in turn are projected from eye-point B into the original rays drawn from A (e.g., point X, which will be a point lying on the conic section curve through A, B, P_1, P_2, P_3). Thus every ray from A corresponds to a ray from B, and the common points of each pair of A- and B-rays all lie on a conic section curve, in this case an ellipse. We have described the construction in detail, and we will not obtain enough points by the above method to draw the ellipse. We can, of course, choose any pair of points from among the five for eye-points and a corresponding point O, so as to obtain the points we require as easily as possible. Fig. 192 shows a whole series of points, and it is clear that they lie on an ellipse. In finding these points we have used four of the "star-points" of the pentagram in turn for the mediating eye-point O because this is more convenient and saves using the lines of the pentagram beyond the star-points. Figs. 193 and 194 show a parabola and hyperbola formed by the same construction. (In these last three drawings, the construction for only a limited number of points is shown. Otherwise there would be too great a confusion of construction lines.)

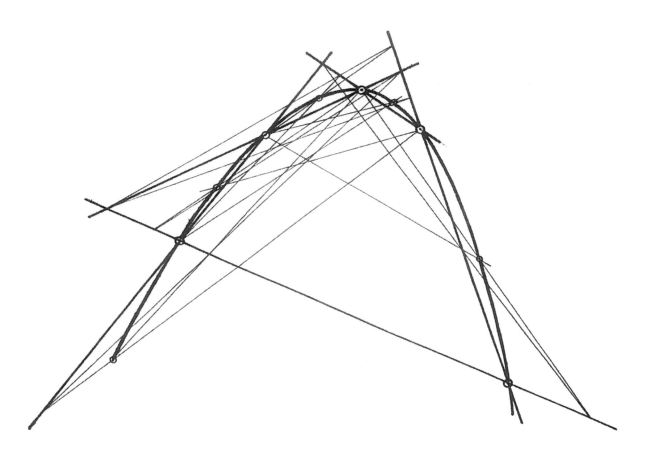

Fig. 193

Five points, then, in a plane (no three of them in a straight line) determine a conic. Under what condition does any sixth point lie on the curve? The answer to this question leads us again to the theorem of Pascal. The addition of a sixth point on the curve determined by the given five points means that we have a hexagon whose opposite pairs of sides must meet in three points lying in a straight line. In Fig. 195, the five points A, B, C, D, E determine the conic, and a sixth point P is added to the curve. We see that the opposite pairs of sides 1,4 and 2,5 and 3,6 meet in three points on the Pascal line, and moreover, one of these points is the mediating eye-point O used in the projective construction of the curve (see Fig. 191). It will be noticed that according to the position chosen for the point P, the Pascal line will pass through one of the five star-points of the pentagram. This leads us to a modified construction for a conic section curve using five random points and five lines determined by these points. The five points are chosen (Fig. 196) and joined by lines 1, 2, 3, 4, and 5. Lines 2 and 5 meet in point (2,5), and lines 1 and 3 meet in point (1,3). These two points determine a Pascal line, p, and line 4 cuts this line p in the point (4,t); the line joining this latter point (4,t) to the point 2 is a tangent to the conic at the point 2 (see Fig. 150a). In a similar manner, the other four tangents may be drawn at points 1, 3, 4, and 5. In our figure, the five Pascal lines are shown in solid line, the five tangents in broken line, and the lines joining the five points in chain line. Thus we have five points and five tangents at these points, and this enables us to draw the conic with considerable accuracy. We may now modify this construction by joining the five random points pentagram-wise instead of pentagon-wise, and this enables us to draw a larger conic in a smaller space (Fig. 197). The chain lines joining the five points pentagram-wise are numbered 1, 2, 3, 4, 5 just as were the lines joining the points pentagon-wise in Fig. 196. The drawing of Fig. 197 is carried out in exactly the same way as that of Fig. 196. We have now shown that a conic section curve may be constructed pointwise by a sequence of projections or perspectives, and by this method we can get as many points as we please, all of them lying on the conic. From this we next developed a method of drawing a conic, with fair accuracy, through five points and touching five tangent lines, the positions and directions of the five lines being dependent on the choice of the five points. Here we have a partly pointwise and partly linewise construction, and this will give us again a further example of the principle of duality. There follows a concise description of the pointwise construction (Fig. 191) and side-by-side with it the same description interchanging the words *point* and *line* (or *ray*). It will be seen that this second description applies to the linewise construction of a conic shown in Fig. 198.

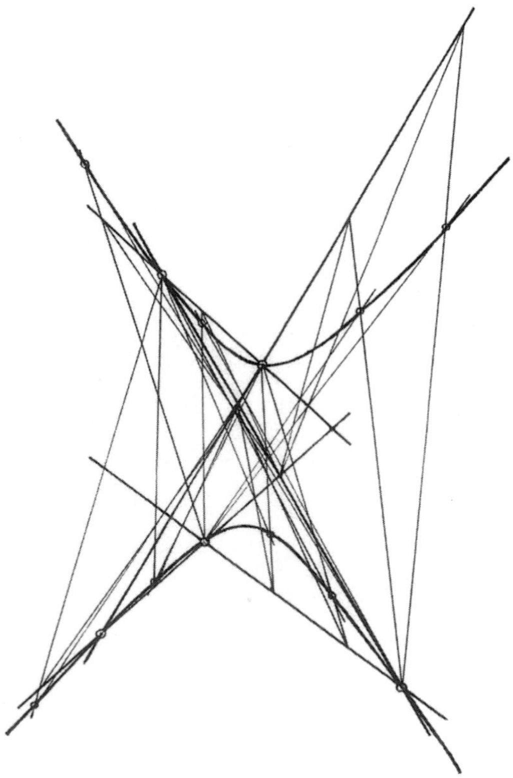

Fig. 194

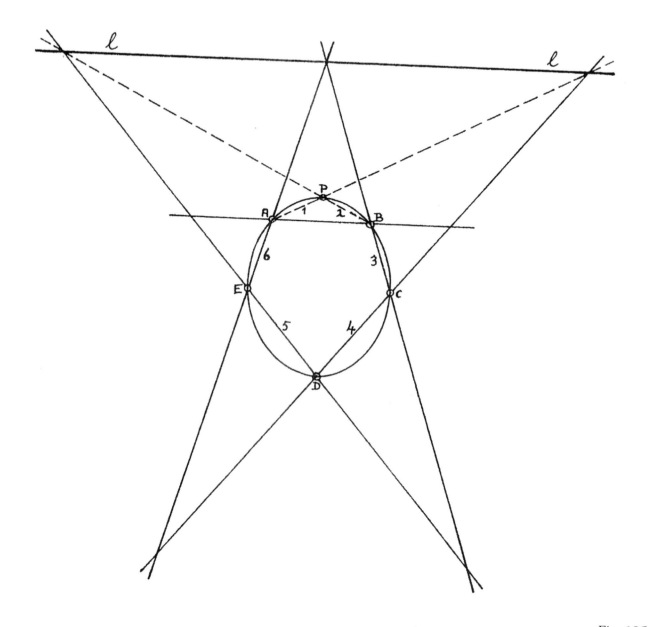

Fig. 195

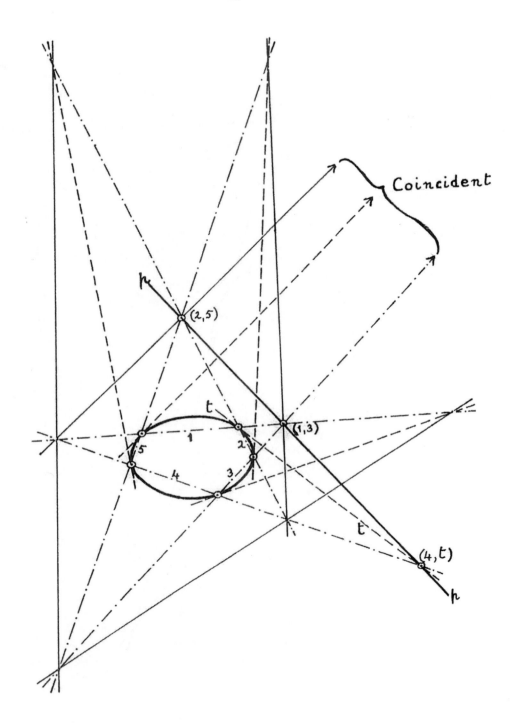

Fig. 196

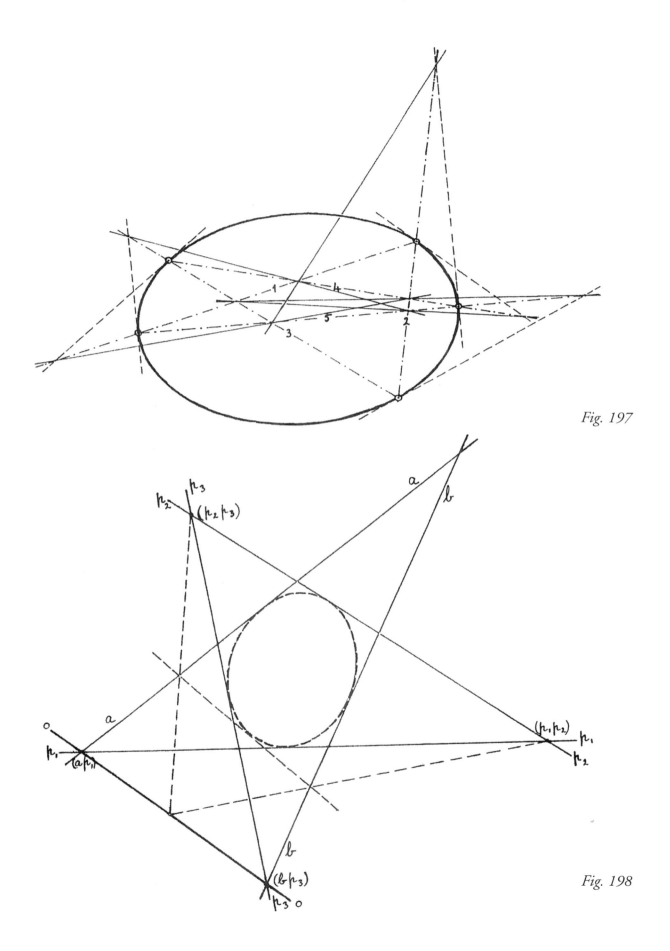

Fig. 197

Fig. 198

Projective Construction of Conic Section Curve through Five Given Points	Projective Construction of Conic Section Curve Using Five Given Lines
Five given points: A, B, P_1, P_2, P_3 chosen as raying points and the other three points P_1, P_2, P_3 serve for fixing the corresponding rays of A and B.	Five given lines: a, b, p_1, p_2, p_3 Two of these lines a and b chosen as ranges of points and the other three lines p_1, p_2, p_3 serve for fixing the corresponding points of a and b.
A-rays give points in line P_2, P_3. These points give rays into the point O, and these rays in turn give points in line P_1, P_2.	Points in line a give rays into point (p_2, p_3). These rays give points into the line o, and these points in turn give rays into the point (p_1, p_2).
Lastly, these latter points give corresponding points in line b.	Lastly, these latter rays give corresponding points in point B.
Thus every A-ray corresponds to a B-ray.	Thus every a-point corresponds to b-point.
The common points of each pair of A- and B-rays all lie on a conic section curve (here an ellipse, Fig. 198).	The common lines of each pair of a- and b-points envelop a conic section curve.
The point O is the mediating point and is the common point of the two lines AP_1 and BP_3.	The line o is the mediating line and is the common line of the two points (ap_1) and (bp_3).
If we had chosen not A and B but two other points as the raying points, we should still have arrived at the same curve.	If we had chosen not a and b but two other lines as a series of giving rays, we should still have arrived at the same curve.

Figs. 199, 200, 201 show the actual molding of ellipse, parabola, and hyperbola by this envelope of lines. Again, as in the case of the pointwise constructions, we can select any pair of the five chosen lines as the series of points through which we draw rays and the corresponding o-line. In the case of the hyperbola drawing (Fig. 201), the two lines crossing between the "wings" of the hyperbola are not the asymptotes.

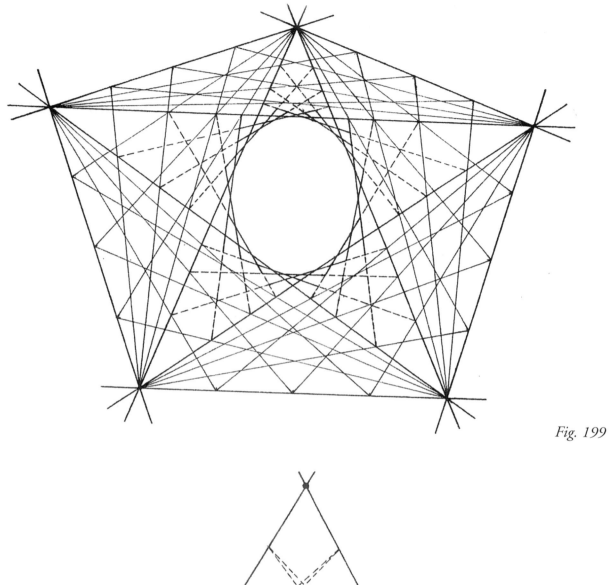

Fig. 199

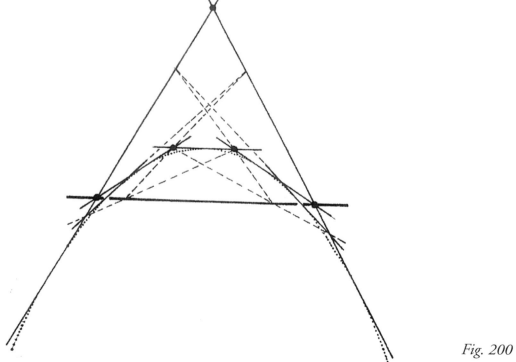

Fig. 200

243

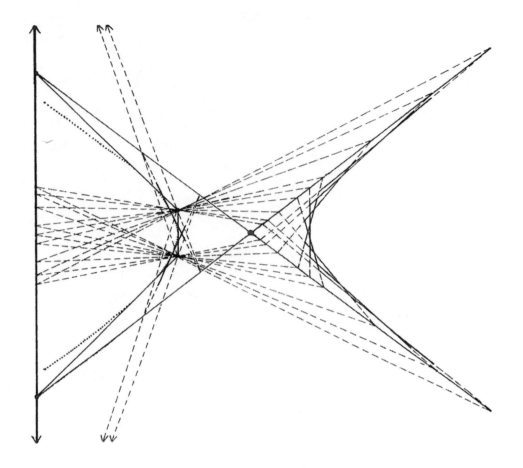

Fig. 201

In these constructions of the conic section curves by sequences of projections, we see how the curves are molded or formed from the outside and the foci are not used, whereas in most of the ordinary constructions we start from the inside of the curve from the foci (see Chapter 7). This molding of forms from the outside belongs essentially to this geometry because projective geometry concerns itself with the pure relationships of points, lines, and planes in space.

In earlier drawings (Figs. 19, 26–27, 56–63, 136), the children experienced this forming of curves by the interweaving of straight lines. It is of real educational value when the teacher can refer to work that has been done perhaps several years ago and show the children the connection between what they did then and the work now in hand. It will be noticed that there are a number of such cross-references in this book.

The fact that the conic section curves arise quite naturally from the pure relationship of lines and points in a plane again reveals to us how fundamental these curves are to our understanding of the nature of space and to the manifold natural forms we see around us.

This chapter is the longest one in this book, but even so only the fringe of the subject has been indicated, and much remains to be done in the educational field in bringing these "new" geometrical ideas into school studies. The author is convinced from his own experience in teaching this subject to older children that it should form an essential part of a school curriculum not merely because it is an interesting and fascinating geometry, but because of its wide cultural value, first in enabling us to gain a fundamental understanding of the world of space and spatial forms, and second, in the wide and manifold application of its forms of thought to the realm of human experience. In recent years, a number of mathematicians in different countries have come to realize the far-reaching importance of this geometry, even recognizing that it is the "geometry of the future." We have already referred to the work of George Adams who, in his recent research, has shown how the forms of living organisms (especially those of the plant kingdom) are to be understood only in relation to projective geometry.

There are indeed indications that projective geometry will be the means by which mathematicians and scientists will come to a new and more fundamental understanding of phenomena of the earth and of the whole universe. In his excellent book *Mathematics in Western Culture* (Geo. Allen & Unwin Ltd., 1954), Morris Kline, towards the end of his final chapter on "Mathematics: Method and Art," writes as follows:

> In any artistic creation the relation of the parts to each other and of the parts to the whole must be harmonious. The harmony in mathematical creations is partly intellectual in the form of logical consistency. The theorems of any one mathematical system must be in complete accord with each other. There are, however, other harmonics. The entire structure of Euclidean geometry is in harmony with the mathematics of number. By means of co-ordinates it is possible to interpret geometrical concepts and theorems algebraically. Conversely, algebraic equations have a geometric interpretation. Thus the two creations are harmonious with each other.

Major mathematical themes have been harmonized with each other. In our brief survey we have touched on four distinct branches of geometry—Euclidean, projective, and two non-Euclidean geometries. As we have viewed these subjects they appear distinct and in some cases inconsistent with each other. Nevertheless, one of the most satisfying mathematical contributions of recent times has shown that it is possible to erect projective geometry on an axiomatic basis in such a way that *the theorems of the other three geometries result as specialized theorems of projective geometry. In other words, the contents of all four geometries are now incorporated in one harmonious whole.*[27]

POSTSCRIPT

During the past 50 years, the methods used in the teaching of geometry have gradually but radically changed. The basic school lessons that I had in this subject during the years of the First World War consisted of learning a considerable number of the propositions of Euclid and their proofs, in much the same form in which they were first enunciated nearly 2,300 years ago, and then solving problems—"riders"—connected with these proofs. This teaching of "classical geometry" changed little during many centuries of education, say, from the monastic schools of the Middle Ages down to quite recent times. As an educational method it was, of course, considered a fine training for the powers of logical thinking because Euclidean geometry itself was regarded as a wonderful edifice built up with the strictest logic on a foundation of simple axiomatic statements. Modern critical research has however revealed that this "structure" is by no means sound, even its foundations being suspect. Thus, strict "Euclid" has largely disappeared from the school curriculum, and the great variety of textbooks that has appeared during the past 20 or 30 years shows how teachers of mathematics are trying to find new methods of presenting geometrical facts and laws. A new edifice is being attempted, built on firmer foundations and using perhaps better materials than those of Euclid. Amid all this change, it is important to notice that the kind of thinking of the architects is unchanged. The forms of thought being used in the teaching of geometry today are fundamentally the same as those inherent in Euclidean geometry. (Here and there, with individual teachers, there are exceptions to this generalization.) Now these forms of thought are just those which Alexis Carrel said must be transformed before we can "undertake the restoration of ourselves and our environment."

In a lecture given in 1919, Rudolf Steiner made the following fateful pronouncement: "Let teaching in our universities go on in the present way for another three decades; let thinking on social matters, as taught today, continue for another thirty years and, at the end of these thirty years, you will find Europe in a state of complete desolation and destruction. . . . Unless a transformation of thinking takes place, Europe will be overwhelmed by a flood of immorality."[28] The reader will notice that 30 years after this lecture was given, Europe was struggling to recover from the devastation and chaos caused by the Second World War.

Many writers and thinkers in various countries have expressed with real conviction the basic problem facing humanity today, the problem is coming up with and exercising a new way of thinking. But few of them tell us what new qualities must come into our thinking, and fewer still tell us how we are to overcome our old habits of thought. Rudolf Steiner has done both these things, and new forms of thought and how we may attain them lie at the foundation of all his teaching. It is not my task in this book to discuss this vital challenge of the transformation of human thinking; I would refer my readers to the works of Rudolf Steiner, especially *The Philosophy of Spiritual Activity* and *The Redemption of Thinking* (see List of Selected Books).

These problems are indeed the concern of all school teachers. It is our present and future task to present our various subjects in the classroom so that the possibility arises for our children to grow into new ways of thinking as they become older—into a thinking that has a fine imaginative quality, that is comprehensive.

I have tried to show that such ways of thinking are inherent in the presentation of geometry as outlined in this book. What I have described are only "stepping stones" towards the path along which the teaching of this subject must surely go in the future, and I am firmly convinced that this will be the path of projective geometry. It is undoubtedly the case that the true logical buildup of this geometry can also supply just this element in education, which the classical Euclidean geometry cannot do. In the last chapter I have suggested how some of the fundamental aspects of projective geometry may be presented to older children. The question now arises if such ideas and thought-forms can be introduced to younger children. I am sure they can, in an indirect way, and I have given some indications of how this may be done. For example, in quite elementary work, children of 12 and 13 can learn to construct curves linewise as well as pointwise (see Chapter 3). In so doing, they are already *experiencing* the polarity of point and line in a plane as well as the nature of some of the fundamental curves that they will study in a more detailed way later. Or again, the drawing of Fig. 36 in Chapter 3 is related to that of Fig. 180 in Chapter 8. By a careful choice of early drawings, much could be done to bring the teaching of this subject into the "one harmonious whole" of projective geometry, and this, I would suggest, is the present and future task of teachers of mathematics.

Endnotes:

1 Published by George Philip & Son, Ltd., 1906.

2 See *Esthetique des Proportions dans la Nature et dans les Arts* by M. C. Gyhka (Gallimard, Paris) and *The Parthenon and Other Greek Temples: Their Dynamic Symmetry* by Jay Hambidge (Yale University Press).

3 *Der Goldene Schnitt* by Dr. Adalbert Goeringer. Munchen 1911, J. Lindauersche Buchandlung.

5 Many Greek and other vases reveal the proportions of the golden ratio. See especially *Dynamic Symmetry, The Greek Vase* by Jay Hambidge (Yale University Press)

6 Quoted by kind permission of the author.

7 Editor's note: Sir Theodore Cook designates the golden ratio proportion by the symbol F; thus F = 1.61803 . . . .

8 *The Curves of Life*, p. 88.

9 See also an article by Joachim Schultz in *Goethe in unsere Zeit*, Dornach, Switzerland.

10 *The Curves of Life*, page 420.

11 See page 130 et. seq. and pages 155, 156.

12 The dihedral angle is the angle between two faces of the figure.

13 This statement appears to be at variance with what we said in the last chapter about the hyperbole being a picture process of subtraction. This contradiction will be resolved in the present chapter.

14 Its highest aspect is the Trinity of the Christian faith, "Three in One and One in Three." But this mystery of the Holy Trinity is also expressed in all Creation—in the body of every human being in body, soul, and spirit—or again in thinking, feeling and willing—in the world of substance, acid, base, and salt, to give only a few examples.

15 E. H. Gombrich, *The Story of Art*, pp. 165, 166. The Phaidon Press Ltd.

16 The author owes a great debt to George Adams for his inspiration and guidance as a teacher, and much of what follows is the result of work done with him.

17 Both here and in what follows the word "line" always means a "straight line" extending to infinity in both directions. Also a plane is always to be thought of as of infinite extent.

18 Some modern writers call it the "principle of polarity."

19 Certain other words have to be changed in some dual statements in order to make sense.

20 From a strictly mathematical point of view, the plane and the point might be regarded as two coincident planes and two coincident points.

21 *Growth and Form* by D'Arcy W. Thompson, pp. 507, 508.

22 Charles Darwin, *The Origin of Species*, Chapter 7, "Cell-Making Instinct of the Hive-Bee."

23 *Growth and Form* by D'Arcy W. Thompson, pp. 538, 539, and 541.

24 We shall use the designation quadrangle where we consider the figure point-wise and quadrilateral when we consider it linewise.

[25] It may also be noted that producing opposite pairs of sides of the original circumscribing quadrilateral also gives two further points lying on the polar (pp).

[26] This harmonic quadrangle is constructed quite at random; that is, the two rays from A and the single ray from P are arbitrarily drawn. We may, however, draw the quadrangle with its angular points lying on the circle (Fig. 173a) formed as before by two rays from A, two from B, and one ray each from P and Q. This drawing is similar to that of Fig. 156. Thus the result of Fig. 173 is implicit in Fig. 156. It should be noted that the four points on the polar line are also harmonic conjugate pairs— x, y and M, y (M being the point where the second diagonal of the quadrangle cuts the polar). The inscribed quadrangle is also the harmonic quadrangle for these latter four points.

[27] Italics added by author.

[28] Italics added by author.

LIST OF SELECTED BOOKS

Rudolf Steiner Education
    The Way of a Child by A. C. Harwood
    Rudolf Steiner Education by L. F. Edmunds
    Childhood: A Study of the Growing Child by Dr. C. von Heydebrand
    The Golden Years by John Benians

By Rudolf Steiner
    Essentials of Education (Five lectures given at the Waldorf School, Stuttgart, 1924)
    Lectures to Teachers (An abridgement of 16 lectures given to teachers in Dornach, Switzerland)
    The Spiritual Ground of Education (Nine lectures given at a "Conference on Spiritual Values in Education and Social Life" at Manchester College, Oxford, 1922)
    The Study of Man (14 lectures given at the Waldorf School in Stuttgart, 1919)
    Education and Modern Spiritual Life (12 lectures given at Ilkley, Yorkshire, 1923)

Mathematics
    Active Arithmetic by Henning Anderson. Fair Oaks, CA: AWSNA Publications, 1995.
    Algebra by Amos Franceschelli. Fair Oaks, CA: AWSNA Publications, 1985.
    Geometry at the Junior High School Grades and *The Waldorf School Plan* by Hermann von Baravalle, Ph.D. Waldorf School, Adelphi College, Garden City, New York, 1948.
    A Rhythmic Approach to Mathematics by Edith L. Somervell. George Philip & Son, Ltd., 1906.
    Das Rechnen im Lichte der Anthroposophie by Ernst Bindel. Waldorf-Spielzeug & Verlag G.M.B.H., Stuttgart.
    "Die Kegelschnitte in menschengemasser Behandlung," article published by Ernst Bindel in *dar naturwissenschaftishen Sektion der Freien Hochschule fur Geisteswissenschaft am Goetheanum*, Dornach, Switzerland.
    Finding the Path: Themes and Methods for the Teaching of Mathematics in a Waldorf School by Bengt Ulin. Fair Oaks, CA: AWSNA Publications, 1991.
    Form Drawing —Grades 1–4, by Ernst Schuberth and Laura Embry-Stein. Fair Oaks, CA: Rudolf Steiner College Press, 2001.
    The Geometry Lesson in the Waldorf School for Classes 4 and 5, Schuberth, Ernst.. Fair Oaks, CA: AWSNA Publications, 2002.
    Geometry for the Waldorf High School by Herbert Swanson. Fair Oaks, CA: AWSNA Publications, 1987.
    Introduction to Advanced Arithmetical Operations for Waldorf School 7th Grades by Ernst Schuberth. Fair Oaks, CA: AWSNA Publications, 2000.
    Math Lessons for the Elementary Grades by Dorothy Harrer. Fair Oaks, CA: AWSNA Publications, 1982.
    Mathematical Models by H. M. Cundy and A. P. Rollet. Clarendon Press, Oxford, 1952.
    Mathematics in the Classroom: Mine Shaft and Sunlight by Amos Franceschelli. Spring Valley, NY: Mercury Press, 1998.
    Mensuration by Amos Franceschelli. Fair Oaks, CA: AWSNA Publications, 1987.
    Projective Geometry by Lawrence Edwards. Edinburgh, Scotland: Floris Books, 1996.

Projective Geometry by Angelo Andes Rovida. Forest Row, England: Steiner Schools Fellowship, 1988.

Projective Geometry by O. Veblen and J. W. Young. Ginn & Co.

Proceedings from the Computer and Information Technology Colloquium, Mitchell, David, ed. Fair Oaks, CA: AWSNA Publications, 2002.

Proceedings from the Mathematics Colloquium by David Mitchell, ed. Fair Oaks, CA: AWSNA Publications, 2001.

Projektive Geometrie by L. Locher-Ernst. Orell Füssli Verlag, Zurich-Leipzig.

Geometry for Advanced Pupils by E. A. Maxwell. Clarendon Press, Oxford.

Solid Geometry: Geometry of the Platonic Solids and the Geometry of the Cylinder, Sphere, and Cone, by Harry Kretz. Fair Oaks, CA: AWSNA Publications, 1999.

Space and Counter Space by Locher-Ernst, Louis. Fair Oaks, CA: AWSNA Publications, 2002.

Space and the Light of the Creation by George Adams Kaufmann. London: A. Renwick Sheen, 1933.

Strahlende Weltgestaltung (Synthetische Geometrie in geisteswissenschaftlicher Beleuchtung) by George Adams Kaufmann. Mathematisch-Astronomische Sektion am Goetheanum, Dornach, Switzerland, 1934.

Teaching Mathematics in Rudolf Steiner Schools by Ronald Jarman. Stroud, Gloucestershire, England: Hawthorn Press, 1998.

Teaching Mathematics for First and Second Grades in Waldorf Schools by Ernst Schuberth. Fair Oaks, CA: Rudolf Steiner College Press, 2000.

The Axioms of Projective Geometry: No. 4 in the Series, Cambridge Tracts in Mathematics and Mathematical Physics (Cambridge University Press).

The Geometry of Form in Nature and in Art

The Curves of Life by Theodore A. Cook (Dover)

Growth and Form by D'Arcy W. Thompson (Cambridge University Press, 1942)

Esthétique des Proportions dans la Nature et dans les Arts by M.C. Ghyka (Gallimard, Paris)

Geometrical Composition and Design by Matila Ghyka (Alec Tiranti Ltd., London)

Der Goldene Schnitt by Dr. Adalbert Goeringer (München 1911, J. Lindauersche Buchandlung)

The Parthenon and Other Greek Temples: Their Dynamic Symmetry by Jay Hambidge (Yale University Press)

Dynamic Symmetry, the Greek Vase by Jay Hambidge (Dover)

The Living Plant by George Adams and Olive Whicher (Goethean Science Foundation, Clent, Stourbridge, Worcester, 1949)

The Plant Between Sun and Earth by George Adams and Olive Whicher (Goethean Science Foundation, Clent, Stourbridge, Worcester, 1952)

The Field of Form by Lawrence Edwards (Floris Books)

Geometrie als Sprache der Formen by Hermann von Baravalle (Verlag Freies Geistesleben)

Miscellaneous

Man, the Unknown by Alexis Carrel (Hamish Hamilton)

Mathematics in Western Culture by Morris Kline (Geo. Allen & Unwin Ltd., 1954)

Eurythmy as Visible Speech by Rudolf Steiner (15 lectures given in Dornach, Switzerland, 1924)

Philosophy of Spiritual Activity by Rudolf Steiner (15 lectures given in Dornach, Switzerland, 1924)

The Redemption of Thinking: A Study in the Philosophy of Thomas Aquinas by Rudolf Steiner
(Three lectures and an epilogue given in 1920)

Some titles may be obtained from:

> AWSNA Publications
> 3911 Bannister Road
> Fair Oaks, CA 95628
>
> 916/961-0927
> http: www.waldorfeducation.org
>
>
> The Anthroposophic Press
> P.O. Box 799
> Great Barrington, MA 01230
>
> 800/856-8664
> http:www.anthropress.org

Index

A

Adams, George 186, 245
Alexander the Great 141
analytical or Cartesian geometry 65
Apollonian circles 121, 123, 224, 227
Apollonius of Perga 123, 142
Arabic numerals 75
Archimedes 141
area of a triangle 60
Aristotle 81
arithmetic and geometric progression 75, 108
Arthurian legends 74
asymptotes 169
axe showing golden ration proportions Fig. 72
axiom of parallelism 186

B

Babylonians 57
bee's cell 199, 200
Bindel, Ernst 173
Book of Chou Pei Suan King 57
Boyle's Law 114, 171
Brianchon, C.J. 200
Brianchon's point 189, 193, 203
Brunelleschi, Filippo 181

C

Carrel, Alexis 24, 247
Carroll, Lewis 167
Cassini, Jean Dominique 115
    Cassini, curves of 115, 124

Cartesian analytical geometry 186
Cartesian coordinates 171
Cartesian geometry 187
Cayley, Arthur 186
Church, Dr. A. H. 109
Churchill, Sir Winston 22
circle 34-35, 62, 65, 193
circle drawing, the fundamental 55
concentric circles 41
conic section curves 49, 171, 173, 193, 236, 238, 245
conic sections 141, 225
continuous decagram band 83
Cook, Sir Theodore 81, 107-109, 115
Copernicus 182
crinoids 60, Fig. 52
cube 125, 130, 133
curve of pursuit 66, 67
curves of addition 172
cylinder 142

D

D'Arcy Thompson 109, 115, 121, 134, 199
Darwin, Charles R. 199
Desargues, Gerard 142, 205 166
    theorem of 208, 217; Figs. 158-67, 174
Desargues' configuration 208, 209, 210
Descartes 186
division treated geometrically Figs. 90, 91
dodecahedron 125, 130, 131, 133; Figs. 92e, 93e, 94f, 98, 99, 100, 102
Donatello 182, Plate 6
Dürer, Albrecht 127, 181; Plate 7

E

echinoderms 60
"economy" theory 200
egg of Cerebratulus 121
Egypt 125
Egyptians 57
elementary geometrical figures 69
ellipse 34, 111, 112, 123, 141, 142, 162, 165, 236; Figs, 85, 100, 105, 110, 115-118, 121, 125, 129

elliptical orbits of planetary motion 126
equilateral 38
equilateral triangle 40
Euclid 57, 141, 166, 247
Euclidean geometry 247, 248
Euclidean geometry, classical 186
Euclidean proofs 69
experience of form 27

F

family of confocal parabolas 154
family of ellipses 112, 154
family of harmonic quadrangles 214
family of hyperbolas 113, 154
family of parabolas 153, 154
Fibonacci, Leonardo 75
Fibonacci angle 108
Fibonacci series 75–80, 83, 87, 108
Findlay, Professor J. J. 33
first school lessons 33
five Platonic solids 128
five regular polyhedra 128
foci 111
form in nature 69
form of the human being 34
free-hand drawing 111
free-hand pattern drawing 28
friendly numbers 82-83

G

Galileo 182
Gauss, Karl Friedrich 75
Gawain and the Green Knight (reference to Pentagram) 74
geometrical laws 60
geometrical progression 45, 80, 108
geometry 33
    Analytical (see Analytical Geometry)
    Euclidean (see Euclid)
    non-Euclidean (see non-Euclidean geometry)
    Projective or Synthetic (see Projective Geometry)
Ginsburg, Dr. J. 78

Giotto 106, Plate 4
Goethe 73
Golden Age of Greek mathematics 141
golden compass 93
golden number 81, 89
golden ratio 68, 88, 89, 90, 93, 94, 95, 97, 103, 106, 107, 108, 109, 110, 134
golden ratio proportions 110
golden rectangle 93, 134; Figs. 74, 75
golden series 91
Goeringer, Dr. 94, 103
Gombrich, E. H. 181
Gothic arches 34
Greek statues 109
Greek temples 94
Greek vase 34
Guidi, Tommaso (Masaccio) 181; Plate 5

H

Haeckel, Ernst 134; Fig. 101
handwork 31
harmonic conjugate pairs 210, 225
harmonic conjugate points 224
harmonic lines 224, 225; Fig. 176 et seq.
harmonic pair 220
harmonic points 210, 216, 225
harmonic quadrangles 210, 217, 225, 230; Figs 168 et seq.
harmonic range 221
Herbart, J. F. 21
hexagon 28, 237; Figs. 52, 137-149, 195
hexaotinellid sponges 134
Hobbema, M. 181
human proportions 110
hyperbola 113, 121, 123, 124, 141, 142, 163, 165, 169, 171, 193, 230; Figs. 145, 146, 183, 187, 194, 201

I

icosahedron 125, 130, 133; Figs. 92d, 93e, 94e, 98, 99, 100
imagination 23, 69 et seq.
infinity 166, 204, 214, 225
inner moral strength 31
inner soul quality of the child 25
intellectual proof 57

isosceles triangles 38
isosceles right-angled triangle 58

K

Keats, John 22, 25
Kepler, Johannes 34, 57, 71, 90, 115, 125, 134, 142, 166
Kine, Morris 23, 245
King Edward II 98

L

law of the hyperbola 170
leaf forms 61
leaves 108
Leonardo da Vinci 90, 103, 109, 181, 182
lily 199
Locher-Ernst, Louis 186
logarithmic spiral 67, 68, 108
logical proof 19
Lord Elton 21

M

Masaccio (Tommaso Guidi) 181; Plate 5
mathematics of the ellipse 169
mathematics of the pentagram 82
median 203
Menaechmus 141
metamorphosis 31, 55, 61, 62, 65, 113, 169
monocotyledon 199
movement 27
movements of the Sun and the stars 173
multiplication 115
mystic hexagram 193

N

nets for making models of the five Platonic solids Fig. 93
Newton, Sir Isaac 182
non-Euclidean geometry 246

O

obtuse angle 85
octahedron 125, 130 et seq.; Figs. 92b, 93b, 94b, 95, 97, 98
On Growth and Form 109
orthogonal projection 133
ovals of Cassini (see Cassini, curves of)

P

Pacioli, L. 90
Pappus 199
parabola 65, 66, 123, 141, 142, 153, 163, 172, 230 et seq.; Figs. 56, 57, 60-63, 104, 106, 109, 112-114, 122, 124, 130, 132, 134, 135
parabolic curve 68, 163
parallelism 166; Fig. 127
parallelogram 49, 50, 55; Figs. 30, 32, 33, 37, 38, 40, 45, 46, 47
Pascal, Blaise 186, 188, 200
Pascal's line 189, 193, 200, 202
Pascal's theorem 199
pentagon 62, 65, 83, 200, 202; Figs. 64-71
pentagram 71, 73, 85, 110; Figs. 64-71
Pheidias 109
Pheidias spiral 107
Phi (θ — the golden number) 108 et seq.
phyllotaxis ratio 108
Plato 23, 125, 141
Platonic bodies 125
Poinsot, Louis 134
polyhedra 125
principle of duality 187
principle of harmonic points 227
principle of the harmonic range 236
projective geometry 57, 186, 187, 200, 220, 236, 248
Pythagoras 57, 60, 81, 93, 125
Pythagorean School of mathematics 81, 93
Pythagorean School of philosophy 71

Q

quadrangular cell 210
quadrilaterals 44, 50, 200, 203; Figs. 40, 41, 46, 157
quartz crystal 199, 200

R

radiolarian 134; Fig. 101
Raphael 182
ratios 76, 87, 108, 220
reciprocal 77
rectangle 44, 50; Figs. 26, 36, 41, 44-47
rectangular hyperbolae 45
relationship between multiplication and addition 124
rhombuses 45, 50
Ridley, M. R. 74
right angles 40, 55

S

scalene triangles 38
scalene right-angled triangle 58
school curricula 69
Schwann, Theodor 199
scientific phenomena 23
sea lilies 60
sea shells 108
space, the nature of 69
Spencer, Herbert 33
spiral 107
spherical aberration 173
spiritual understanding of phenomena 171
square 44, 55; Figs. 20-29, 42, 114
St. Francis preaching to the birds 106
starfish 60
Steiner, Jacob 186
Steiner, Rudolf 186, 247, 248
structure of common minerals 134
Somervell, Mrs. Edith L. 25, 65
symmetry 28, 134, 224; Figs. 1, 1a, 1b
symmetry of the human form 153

T

tetrahedron 125, 130 et seq.; Figs. 92c, 93c, 94c, 94d, 96-98
The Last Supper 103, 110; Plate 3
theorem of Pythagoras 57, 71
theorem of Desargues 205, 217
theorem of pole and polar 201

thinking, feeling, and willing 22
training of the child's intellect 21
transcendental curve (see "Curve of Pursuit")
trapezoid 49; Fig. 31, 38, 39, 47
triads of Menaechmus 141
triangles 28, 38, 40, 44, 65, 203; Figs. 8-19, 49-51
Trine, Ralph Waldo 205
truth 19
Tycho de Brahe 126

U

universal phenomenon 201

V

vase showing golden ratio proportions Fig. 80
vectorial angle 108
violin showing golden ratio proportions 94, 95; Figs. 78, 79
visible speech 27
von Baravalle, Dr. Hermann 65

W

Walter, the King's Painter (maker of the Coronation Chair) 97, 98; Figs. 82, 83
water-lily leaf 62
Whitehead, Alfred North 187